Theory and Applications
of Natural Language Processing

Aims and Scope

The field of Natural Language Processing (NLP) has expanded explosively over the past decade: growing bodies of available data, novel fields of applications, emerging areas and new connections to neighboring fields have all led to increasing output and to diversification of research.

"Theory and Applications of Natural Language Processing" is a series of volumes dedicated to selected topics in NLP and Language Technology. It focuses on the most recent advances in all areas of the computational modeling and processing of speech and text across languages and domains. Due to the rapid pace of development, the diversity of approaches and application scenarios are scattered in an ever-growing mass of conference proceedings, making entry into the field difficult for both students and potential users. Volumes in the series facilitate this first step and can be used as a teaching aid, advanced-level information resource or a point of reference.

The series encourages the submission of research monographs, contributed volumes and surveys, lecture notes and textbooks covering research frontiers on all relevant topics, offering a platform for the rapid publication of cutting-edge research as well as for comprehensive monographs that cover the full range of research on specific problem areas.

The topics include applications of NLP techniques to gain insights into the use and functioning of language, as well as the use of language technology in applications that enable communication, knowledge management and discovery such as natural language generation, information retrieval, question-answering, machine translation, localization and related fields.

The books are available in printed and electronic (e-book) form:

* Downloadable on your PC, e-reader or iPad
* Enhanced by Electronic Supplementary Material, such as algorithms, demonstrations, software, images and videos
* Available online within an extensive network of academic and corporate R&D libraries worldwide
* Never out of print thanks to innovative print-on-demand services
* Competitively priced print editions for eBook customers thanks to MyCopy service http://www.springer.com/librarians/e-content/mycopy

For further volumes:
http://www.springer.com/series/8899

Chris Biemann

Structure Discovery in Natural Language

Foreword by Antal van den Bosch

Chris Biemann
Computer Science Department
Technische Universität Darmstadt
Hochschulstr. 10
64289 Darmstadt
Germany

Foreword by
Antal van den Bosch
Centre for Language Studies
Radboud University Nijmegen
P.O. Box 9103
6500 HD Nijmegen
The Netherlands

ISSN 2192-032X e-ISSN 2192-0338
ISBN 978-3-642-25922-7 e-ISBN 978-3-642-25923-4
DOI 10.1007/978-3-642-25923-4
Springer Heidelberg Dordrecht London New York

Library of Congress Control Number: 2011944202

Mathematics Subject Classification (2010): 68T50, 91F20, 05C82, 62H30, 68T05

Printed on acid-free paper

Springer is part of Springer Science+Business Media (www.springer.com)

Dedicated to my parents, who unsupervised me in a good way

Foreword

Few topics in computational linguistics catch the imagination more than unsupervised language learning. 'Unsupervised' has a magical ring to it. Who would not want to have a system do all the work automatically, and knowledge-free? I suspect it is a common experience of researchers in the field, typically occurring while staring out of a window taking a sip of coffee, to feel a sudden exhilaration: seeing the possibility of circumventing the vexing bottleneck of annotated data gathering with an unsupervised learning algorithm and a lot of unannotated data.

Let us pause here for a bit of exegesis. What does it mean for computational language learning to be *unsupervised*? As just suggested, it involves learning some yet unspecified language processing task on the basis of *unannotated* linguistic data. This in turn begs the question what it means for linguistic data to be unannotated; a reasonable answer would be that unannotated data only consists of linguistic surface elements: sounds or letters, and is devoid of any abstract linguistic elements. This means that any linguistic theory that assumes abstract linguistic elements, be it part-of-speech tags, syllabic roots, or syntactic dependencies, will not be playing any role in unsupervised language learning. This is quite a provocative proposition, and another cause for the rebellious allure of unsupervised learning.

The word *unsupervised*, to continue my exegesis, strengthens the anti-authoritarian connotation even more, but also triggers the question what or whose supervision is thrown overboard. Are we talking about the poor linguist whose wise lessons are ignored? This indeed seems to be the answer suggested by this book, where within a few pages you will read that "Unsupervised means that no labelled training material is provided as input. The machine is exposed to language only, without being told what its output should look like." (this book, p. 2). This is linked to the concept of *knowledge-free*, which is taken to mean "that no knowledge about the specific language, such as e.g. word order constraints or a list of personal pronouns, is given to the system." (this book, p. 2).

Thus, unsupervised language learning is learning from unlabeled data, where labels denote abstract linguistic notions. This aligns well with the parallel meaning attributed to unsupervised learning in the field of machine learning: learning without classification labels. But is the distinction between abstract and surface linguistic elements always clear-cut? Counter to the usual assumption, I argue it is not. Consider, for example, the function word *the*. What is the difference of saying this word carries a syntactic function, and saying that the word *is* a syntactic function? In *the book*, the word *the* marks the beginning of a noun phrase. In *we book*, the word *we* signals a main verb coming up that refers to the first person plural.

If we accept the proposition that linguistic elements such as letters and words can be seen as labels themselves, it is easy to see that supervised machine learning algorithms could be applied to classify strings of linguistic elements into other elements or strings of elements. Learning is still knowledge-free and devoid of linguistic abstractions, but is it unsupervised? If it is not unsupervised in the machine learning sense, can it still be called unsupervised in the no-linguistic-abstractions sense? I would like to disagree, and instead call this type of learning *autosupervised*.

Autosupervised language learning does not occur in an obscure corner of pathological language task definitions. Rather, it is the type of learning that occurs in most present-day statistical text-to-text processing systems: machine translation, paraphrasing, and spelling correction. Even *n*-gram language models with back-off smoothing can be considered to be the product of a simple self-supervised learning procedure producing a decision list or tree that predicts the next word given the previous $n-1$ words.

Allow me to continue, at least in this Foreword, the use of the word *autosupervised* where normally you would have read *unsupervised*.

Our understanding and command of autosupervised language learning, though as a scientific endeavour still rather young and perhaps just out of its infancy, has deepened over the past two decades. Its potential has in fact become one of the key research questions in computational linguistics. We can now build on the shoulders of pioneers such as Hinrich Schütze, Steve Finch and Nick Chater, Ramon Ferrer-i-Cancho, Chris Manning, Dan Klein, and Alex Clark, and before you lies an important next step in this increasing body of work.

While in computational linguistics the topic is close to reaching the grail status that topics such as machine translation have, it is stereotypical for the field to be largely oblivious to theories from other fields that work with the same idea of data-driven discovery of models of language. These are not *computational* models, but the articles and books in which they are described provide a wealth of inspiration, also in hindsight, for the development of computational models of autosupervised language learning. Fortunately, the better work in the area does acknowledge its roots in 20th-century linguistics, with proponents such as J.R. Firth and Zellig Harris, and occasionally also points to the work of present-day usage-based linguists such as Robert Langacker, William Croft, and Adele Goldberg, and the developmental psychologist Michael Tomasello.

It is important to realize that the boundary between non-computational models of autosupervised language learning and their computational counterparts is thin, and could become void if both sides would work with the same concepts and formalizations. Usage-based linguistic theories have been occasionally criticized for not being entirely formal, but I am convinced it is only a matter of time before this gap will be bridged, thanks to work from both sides of the divide in overlap areas such as corpus linguistics. On the one side one finds studies such as the collostructional analysis work of Stefanowitsch and Gries [229]. On the other side one finds the type of work in autosupervised language learning exemplified by this book.

What would be the formal basis that would connect and equalize work from both sides of the divide? Autosupervised language learning methods have tended to build on light formalizations that make use of simple spaces and metrics (vector spaces, bags of words, n-gram models). Despite their sobering simplicity and their complete implicitness, these models are known to harbour the incredible strength that fuels the world's leading search engines, speech recognizers and machine translation systems. Usage-based linguistics, on the other hand, assumes structures (with names such as constructions, collostructions, complex lexical items) of mildly higher complexity: they may have gaps, required and optional elements, and relations between these elements that signal inequality (e.g. dependence). Where bags of words and the Markovian assumption have no answer to these requirements, graph theory does. The book you are now reading describes the building of a machine that starts with this assumption.

This book is built around the concept of an autosupervised structure discovery machine that discovers structure in language data, and can do so iteratively, so that it can discover structure in structured data. It is shown to grasp language identification, part-of-speech tagging, and lexical substitution to levels that rival supervised approaches. To leave sufficient suspense, I trust you will be thrilled to read how.

Nijmegen, September 2011 *Antal van den Bosch*

Preface

After 60 years of attempts to implement natural language competence in machines, there is still no automatic language processing system that comes even close to human language performance.

The fields of Computational Linguistics and Natural Language Processing predominantly sought to teach machines a variety of subtasks of language understanding either by explicitly stating processing rules or by providing annotations they should learn to reproduce. In contrast to this, *human* language acquisition largely happens in an unsupervised way — the mere exposure to numerous language samples triggers acquisition processes that imprint the generalisation and abstraction needed for understanding and speaking that language.

Exactly this strategy is pursued in this work: rather than telling machines how to process language, one instructs them how to discover structural regularities in text corpora. Shifting the workload from specifying rule-based systems or manually annotating text to creating processes that employ and utilise structure in language, one builds an inventory of mechanisms that — once they have been verified on a number of datasets and applications — are universal in a way that allows their application to unseen data with similar structure. This enormous alleviation of what is called the "acquisition bottleneck of language processing" gives rise to a unified treatment of language data and provides accelerated access to this part of our cultural memory.

Now that computing power and storage capacities have reached a sufficient level for this undertaking, we for the first time find ourselves able to leave the bulk of the work to machines and to overcome data sparseness by simply processing larger batches of data.

In Chapter 1, the *Structure Discovery* paradigm for Natural Language Processing is introduced. This is a framework for learning structural regularities from large samples of text data, and for making these regularities explicit by introducing them in the data via self-annotation. In contrast to the predominant paradigms, Structure Discovery involves neither language-specific knowledge nor supervision and is therefore independent of language, domain and data representation. Working in this paradigm instead means establishing procedures that operate on raw language material and iteratively enrich the data by using the annotations of previously applied Structure Discovery processes. Structure Discovery is motivated and justified by discussing this paradigm along Chomsky's levels of adequacy for linguistic theories. Further, the vision of the complete Structure Discovery Machine is outlined: a series of processes that make it possible to analyse language data by proceeding from the generic to the specific. Here, abstractions of previous processes are used to discover and annotate even higher abstractions. Aiming solely at identifying structure, the effectiveness of these processes is judged by their utility for other processes that access their annotations and by measuring their contribution in application-based settings. A data-driven approach is also advocated on the side of defining these applications, proposing crowdsourcing and user logs as means to widen the data acquisition bottleneck.

Since graphs are used as a natural and intuitive representation for language data in this work, Chapter 2 provides basic definitions of graph theory. As graphs based on natural language data often exhibit scale-free degree distributions and the Small World property, a number of random graph models that also produce these characteristics are reviewed and contrasted along global properties of their generated graphs. These include power-law exponents approximating the degree distributions, average shortest path length, clustering coefficient and transitivity.

When defining discovery procedures for language data, it is crucial to be aware of quantitative language universals. In Chapter 3, Zipf's law and other quantitative distributions following power laws are measured for text corpora of different languages. The notion of word co-occurrence leads to co-occurrence graphs, which belong to the class of scale-free Small World networks. The examination of their characteristics and their comparison to the random graph models as discussed in Chapter 2 reveals that none of the existing models can produce graphs with degree distributions found in word co-occurrence networks.

For this a generative model is needed, which accounts for the property of language being a time-linear sequence of symbols, among other things. Since previous random text models fail to explain a number of characteristics and distributions of natural language, a new random text model is developed, which introduces the notion of sentences in a random text and generates sequences of words with a higher probability, the more often they have been generated before. A comparison with natural language text reveals that this model successfully explains a number of distributions and local word order restrictions in a fully emergent way. Also, the co-occurrence graphs of its random corpora comply with the characteristics of their natural language counterparts. Due to its simplicity, it provides a plausible explanation for the origin of these language universals without assuming any notion of syntax or semantics.

In order to discover structure in an unsupervised way, language items have to be related via similarity measures. Clustering methods serve as a means to group them into clusters, which realises abstraction and generalisation. Chapter 4 reviews clustering in general and graph clustering in particular. A new algorithm, Chinese Whispers graph partitioning, is described and evaluated in detail. At the cost of being non-deterministic and formally not converging, this randomised and parameter-free algorithm is very efficient and particularly suited for Small World graphs. This allows its application to graphs of several million vertices and edges, which is intractable for most other graph clustering algorithms. Chinese Whispers is parameter-free and finds the number of parts on its own, making brittle tuning obsolete. Modifications for quasi-determinism and possibilities for obtaining a hierarchical clustering instead of a flat partition are discussed and exemplified. Throughout this work, Chinese Whispers is used to solve a number of language processing tasks.

Chapters 5–7 constitute the practical part of this work: Structure Discovery processes for Natural Language Processing using graph representations.

First, a solution for sorting multilingual corpora into monolingual parts is presented in Chapter 5, involving the partitioning of a multilingual word co-occurrence graph. The method has shown to be robust against a skewed distribution of the sizes of monolingual parts and is able to distinguish between all but the most similar language pairs. Performance levels comparable to trained language identification are obtained without providing training material or a preset number of involved languages.

In Chapter 6, an unsupervised part-of-speech tagger is constructed, which induces word classes from a text corpus and uses these categories to assign word classes to all tokens in the text. In contrast to previous attempts, the method introduced here is capable of building significantly larger lexicons, which results in higher text coverage and therefore more consistent tagging. The tagger is evaluated against manually tagged corpora and tested in an application-based way. The results of these experiments suggest that the benefits of using this unsupervised tagger or a traditional supervised tagger are equal for most applications, rendering unnecessary the tremendous annotation efforts involved in creating a tagger for a new language or domain.

The problem of word sense ambiguity is discussed in detail in Chapter 7. A Structure Discovery process is set up, which is used as a feature to successfully improve a supervised word sense disambiguation system. On this basis, a high-precision system for automatically providing lexical substitutions is constructed.

The conclusion in Chapter 8 may be summarised as follows: Unsupervised and knowledge-free Natural Language Processing in the Structure Discovery paradigm has proven to be successful and capable of producing a processing quality equal to that of conventional systems, assuming that sufficient raw text can be provided for the target language or domain. It is therefore not only a viable alternative for languages with scarce annotated resources, but also overcomes the acquisition bottleneck of language processing for new tasks and applications.

Darmstadt, November 2011 *Chris Biemann*

Contents

Acronyms

Lists of abbreviations used frequently in this volume:

BA	Barabási-Albert Model: a process to generate scale-free graphs
BNC	British National Corpus: a collection of 100 million tokens of British English from different genres
CL	Computational Linguistics: the research area of building linguistic applications with computers
CRF	Conditional Random Field: a supervised machine learning classifier, commonly used for sequence tagging
CW	Chinese Whispers graph clustering algorithm, introduced in this work
DM	Dorogovtsev-Mendes Model: a process to generate scale-free small world graphs with two power law regimes
EP	Entropy Precision: a measure to compare clusterings
ER	Erdős-Rényi Model: a random graph model
F	F-measure: The harmonic mean between precision P and recall R
LCC	Leipzig Corpora Collection: a collection of plain text corpora of standardized size for a large number of languages
LDA	Latent Dirichlet Allocation: a generative clustering algorithm of the topic model family
LSA	Latent Semantic Analysis: a vector space transformation method based on Singular Value Decomposition
MCL	Markov Chain Clustering: a graph clustering method based on random walks
MFS	Most Frequent Sense: strategy of assigning the most frequent sense in WSD, commonly used as a baseline system
MI	Mutual Information: a measure for the dependence between random variables
NER	Named Entity Recognition: the task of finding names in natural language text
NLP	Natural Language Processing: the research area of building systems that can process natural language material

nMI	normalised Mutual Information: the normalised variant of MI
OOV	Out Of Vocabulary rate: the percentage of tokens in a text not known to the model
P	Precision: the number of correct answers divided by the number of total answers
POS	Parts Of Speech: syntactic word classes like verb, noun, pronoun
R	Recall: the number of correct answers given divided by the number of correct answers possible
SD	Structure Discovery: the process of finding regularities in data and annotating them back into the data for later processes
SDM	Structure Discovery Machine: a stack of Structure Discovery processes
ST	Steyvers-Tenenbaum model: a process to generate scale-free small world graphs
SVD	Singular Value Decomposition: a matrix factorisation
SWG	Small World Graph: a graph with a high clustering coefficient and short average path lengths
TWSI	Turk bootstrap Word Sense Inventory: an alternative word sense inventory based on crowdsourcing
WS	Watts-Strogatz-model: a process to generate small world graphs
WSI	Word Sense Induction: task of identifying different meanings of a word
WSD	Word Sense Disambiguation: the assignment of one out of several possible word meanings for a word in context

Chapter 1
Introduction

"Linguistics accordingly works continuously with concepts forged by grammarians without knowing whether or not the concepts actually correspond to the constituents of the system of language. But how can we find out? And if they are phantoms, what realities can we place in opposition to them?" — de Saussure [211, p. 110]

Abstract The Structure Discovery paradigm for Natural Language Processing is introduced. This is a framework for learning structural regularities from large samples of text data, and for making these regularities explicit by introducing them in the data via self-annotation. In contrast to the predominant paradigms, Structure Discovery involves neither language-specific knowledge nor supervision and is therefore independent of language, domain and data representation. Working in this paradigm means to set up discovery procedures that operate on raw language material and iteratively enrich the data by using the annotations of previously applied Structure Discovery processes. Structure Discovery is motivated and justified by discussing this paradigm along Chomsky's levels of adequacy for linguistic theories. Further, the vision of the complete Structure Discovery Machine is sketched: A series of processes that allow analysing language data by proceeding from generic to specific. At this, abstractions of previous processes are used to discover and annotate even higher abstractions. Aiming solely to identify structure, the effectiveness of these processes is judged by their utility for other processes that access their annotations and by measuring their contribution in application-based settings. A data-driven approach is also advocated when defining these applications, proposing crowdsourcing and user logs as a means to widen the data acquisition bottleneck.

1.1 Structure Discovery for Language Processing

In the past, Natural Language Processing (NLP) has always been based on *explicit* or *implicit* use of linguistic knowledge. Explicit rule based approaches prevail in classical linguistic applications, while machine learning algorithms use implicit knowledge for generating linguistic annotations.

The question behind this work is: how far can we go in NLP without assuming any linguistic knowledge? How much effort in annotation and resource building is needed for what level of sophistication in natural language analysis?

Working in what I call the *Structure Discovery* (SD) paradigm, the claim being made here is that the required knowledge can largely be acquired by knowledge-free and unsupervised methods. By employing knowledge about language universals (cf. Chapter 3) it is possible to construct Structure Discovery processes, which operate on a raw text corpus[1] in an iterative fashion to unveil structural regularities in the text and to make them explicit in a way that further SD processes can build upon it.

A frequent criticism on work dealing with unsupervised methods in NLP is the question: "Why not take linguistic knowledge into account?" The simple answer to this is that for many languages and applications, the appropriate linguistic knowledge just is not available. While annotated corpora, classification examples, sets of rules and lexical semantic word nets of high coverage exist for English, this does not reflect the situation for most of major world languages. Further, handmade and generic resources often do not fit the application domain, whereas resources created *from and for* the target data inherently do not suffer from these discrepancies.

Structure in this context is any kind of automatically acquired annotation that relates elements of language along arbitrary kinds of similarity. It is not restricted to a single language level, but encompasses labels with the scope of e.g. sentences, clauses, words or substrings. Since the processes that discover and mark structural regularities are not allowed to choose from a predefined inventory of labels, the names of the labels are meaningless and receive their interpretation merely through the elements marked by them.

Unlike other works that proceed by teaching the machine directly how to solve certain tasks — be it by providing explicit rules or implicitly by training the machine on handcrafted examples — the scope of this work is the unsupervised, knowledge-free acquisition of structural regularities in language. Unsupervised means that no labelled training material is provided as input. The machine is exposed to language only, without being told what its output should look like. Knowledge-free means that no knowledge about the specific language, such as e.g. word order constraints or a list of personal pronouns, is given to the system. The work of a Structural Discovery engineer is rather to provide a suitable collection of natural language data and to set up procedures that make use of it.

All knowledge about how to conduct this augmentation of structure is encoded procedurally in methods and algorithms. This keeps the paradigm entirely language independent and applicable without further effort to all human languages and sub-languages for which data is available. Given the fact that several thousand human languages are spoken in the world and given the ongoing specialisation in science, production and culture, which is reflected in respective sub-languages or domains, this paradigm provides a cheaper, if not the only way of efficiently dealing with this variety.

[1] the terms text, text data, corpus, text corpus, language and language data are used interchangeably throughout this work.

Shifting the workload from creating rich resources manually to developing generic, automatic methods, a one-size-fits-all solution needing only minimal adaptation to new domains and other languages comes into reach. While the final task is defined by the application, and data has still to be collected that defines the target behaviour of the system, the core point of Structure Discovery is to fully automatise the natural language preprocessing steps that are necessary to build the application.

In the remainder of this section, the paradigms followed by the fields of Computational Linguistics (CL) and Natural Language Processing (NLP) are shortly contrasted after laying out the SD paradigm in more detail. Further, the benefits and drawbacks of knowledge-free as opposed to knowledge-intensive approaches, as well as degrees of supervision are elaborated on and the SD paradigm is compared to other paradigms.

1.1.1 Structure Discovery Paradigm

The Structure Discovery (SD) paradigm is the research line of algorithmic descriptions that find and employ structural regularities in natural language data. The goal of SD is to enable machines to grasp regularities, which are manifold and necessary in data used for communication, politics, entertainment, science and philosophy, purely from applying operationalised procedures on data samples. Structural regularities, once discovered and marked as such in the data, allow an abstraction process by subsuming structural similarities of basic elements. These complex bricks constitute the building block material for meta-regularities, which may themselves be subject of further exploration.

The iterative methodology of the SD paradigm is laid out in Figure 1.1. It must be stressed that working in the SD paradigm means to proceed in two directions: using the raw data for arriving at generalisations and using these generalisations to structurally enrich the data. While the first direction is conducted by all clustering approaches, the second direction of feeding the results back into the data to perform self-annotation is only rarely encountered.

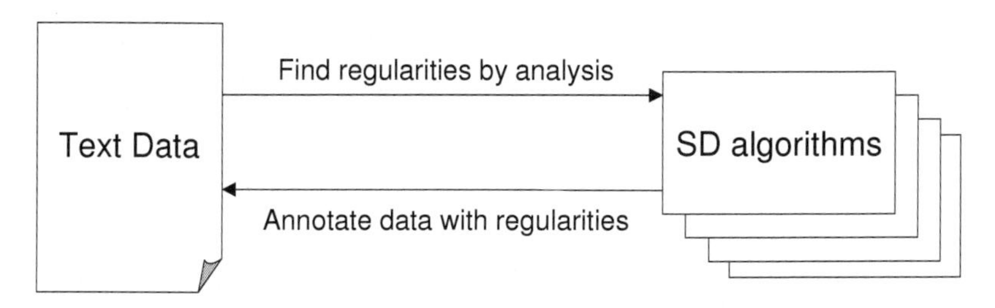

Fig. 1.1 Iterative methodology of the Structure Discovery paradigm: SD algorithms find and annotate regularities in text data, which can serve as input for further SD algorithms

Unlike other schools that provide knowledge of some sort to realise language processing, a system following the SD paradigm must arrive at an adequate enrichment of language data with the instances of the discovered structures, realising self-annotation of the data, as merely the data in its raw form determines the kind of structures found and instances identified thereof.

Solely working on algorithmically discoverable structures means to be consequently agnostic to linguistic theories. It is not possible for machines to create proposals for analysis based on intuition and introspection. Rather, the grammarian's knowledge can stimulate discovery process formulation. The native speaker's intuition about particular languages is replaced by a theory-independent intuition of how to discover structure. While as well created intellectually, the utility of the new discovery process can immediately be judged and measured. Further, it is applicable for all data exhibiting similar structure and thus more general.

1.1.2 Approaches to Automatic Language Processing

The discipline of automatically processing natural language is split into two well-established subfields that are aimed at slightly different goals. Computational Linguistics (CL) is mainly influenced by linguistic thinking and aims at implementing linguistic theories following a *rationalist* approach. Focus is set on linguistic problems rather than on robust and efficient processing. Natural Language Processing (statistical NLP, in the definition of Manning and Schütze [158]), on the other hand, is not linked to any linguistic theory. The goal of NLP is not understanding of language structure as an end in itself. Rather, knowledge of language structure is used to build methods that help in processing language data. The criterion of NLP is the performance of the system, not the adequacy of the representation to human language processing, taking a pragmatic but theoretically poorly founded view. Regarding current research, the two fields are not easily separated, as they influence and fertilise each other, so there is rather a continuum than a sharp cutting edge between the two.

The history of automatic treatment of natural languages was dominated consecutively by two major paradigms: rule-based and statistical approaches. Starting with the realm of computers, rule-based approaches tackled more and more problems of language processing. The leitmotif was: given enough time to develop more and more sophisticated rules, eventually all phenomena of language will be encoded. In order to operationalise the application of grammar formalisms, large systems with several thousand grammar rules were built. Since these rules interact with each other, the process of adding sensible rules gets slower the more rules are already present, which makes the construction of rule-based systems expensive. What is inherent of the top-down, introspection-driven construction of rule-based systems is that they can only work on the subset of language covered by the rules. History has shown that the size of this subset remained far from getting close to full coverage, resulting in only a fraction of sentences being processable.

Since the advent of large machine-readable text corpora starting with the Brown Corpus [103] and the computing power necessary to handle them, statistical approaches to language processing received increased interest. By basing decisions on probabilistic models instead of rules, this *empiricist* approach early showed to be capable of reaching higher accuracy on language processing tasks than the rule-based approach. The manual labour in statistical methods is shifted from instructing the machines directly by rules how to process the data to labelling training examples that provide information on how a system's output should look like. At this point, more training means a richer basis for the induction of probabilistic models, which in turn leads to better performance.

For understanding language from a linguistic point of view and testing grammar theories, there does not seem to be another way than the rule-based approach. For building applications, however, statistical methods proved to be more robust and to scale better, which is probably best contrasted for Machine Translation by opposing Martin Kay's essay "Machine Translation: The Disappointing Past and Present" [137] to Franz Josef Och's talk "Statistical Machine Translation: The Fabulous Present and Future" [187].

Completely opposed to the rule-based approach, the work described in this volume takes the statistical approach a step further by not even allowing implicit knowledge to be provided for training.

1.1.3 Knowledge-intensive and Knowledge-free

Another dimension, along which it is possible to classify language processing methods, is the distinction between knowledge-intensive and knowledge-free approaches, see also [42]. Knowledge-intensive approaches make excessive use of language resources such as dictionaries, phrase lists, terminological resources, name gazetteers, lexical-semantic networks such as WordNet [178], thesauri such as Roget's Thesaurus [204], ontologies and the like. As these resources are necessarily incomplete (all resources leak), their benefit will only extend to a certain point. Additional coverage can only be reached by substantial enlargement of the resource, which is often too much of a manual effort.

But knowledge-intensiveness is not only restricted to explicit resources: the rules in a rule-based system constitute a considerable amount of knowledge, just like positive and negative examples in machine learning.

Knowledge-free methods seek to eliminate human effort and intervention. The human effort is not in specifying rules or examples, but in the method itself, lending the know-how by providing discovery procedures rather than presenting the knowledge itself. This makes knowledge-free methods more adaptive to other languages or domains, overcoming the brittleness of knowledge-intensive systems when exposed to an input substantially different from what they were originally designed for.

Like above, there rather is a continuum than a sharp border between the two ends of the scale. Methods that incorporate only little human intervention are sometimes labelled knowledge-weak, combining the benefits of not having to prepare too much knowledge with obtaining good results by using available resources.

1.1.4 Degrees of Supervision

Another classification of methods, which is heavily related to the amount of knowledge, is the distinction between supervised, semi-supervised, weakly supervised and unsupervised methods.

- In *supervised* systems, the data as presented to a machine learning algorithm is fully labelled. That means: all examples are presented with a classification that the machine is meant to reproduce. For this, a classifier is learned from the data, the process of assigning labels to yet unseen instances is called classification.
- In *semi-supervised* systems, the machine is allowed to additionally take unlabelled data into account. Due to a larger data basis, semi-supervised systems often outperform their supervised counterparts using the same labelled examples (see [254] for a survey on semi-supervised methods and [210] for a summary regarding NLP and semi-supervision). The reason for this improvement is that more unlabelled data enables the system to model the inherent structure of the data more accurately.
- Bootstrapping, also called self-training, is a form of learning that is designed to use even less training examples, therefore sometimes called *weakly-supervised*. Bootstrapping (see [23] for an introduction) starts with a few training examples, trains a classifier, and uses thought-to-be positive examples as yielded by this classifier for retraining. As the set of training examples grows, the classifier improves, provided that not too many negative examples are misclassified as positive, which could lead to deterioration of performance.
- *Unsupervised* systems are not provided any training examples at all and conduct clustering. This is the division of data instances into several groups, (see [158, Ch. 14] and [20] for an overview of clustering methods in NLP). The results of clustering algorithms are data driven, hence more 'natural' and better suited to the underlying structure of the data. This advantage is also its major drawback: without a possibility to tell the machine what to do (like in classification), it is difficult to judge the quality of clustering results in a conclusive way. But the absence of training example preparation makes the unsupervised paradigm very appealing.

To elaborate on the differences between knowledge-free and unsupervised methods, consider the example of what is called unsupervised (also called knowledge-based) word sense disambiguation [cf. 251]. Word sense disambiguation (WSD) is the process of assigning one of many possible senses to ambiguous words in the text. This can be done supervisedly by learning from manually tagged examples. In the terminology of Senseval-3 [172], an unsupervised WSD system decides the word senses merely based on overlap scores of the word's context and a resource containing word sense definitions, e.g. a dictionary or WordNet. Such a system is unsupervised, as it does not require training examples, but knowledge-intensive due to the provision of the lexicographic resource.

1.1.5 Contrasting Structure Discovery with Previous Approaches

To contrast the paradigm followed by this work with the two predominant paradigms of using explicit or implicit knowledge, Table 1.1 summarises their main characteristics.

Paradigm	classical CL	statistical NLP	SD
Approach	rule-based	statistics	statistics
Direction	top-down	bottom-up	bottom-up
Knowledge Source	manual resources	manual annotation	–
Knowledge Intensity	knowledge-intensive	knowledge-intensive	knowledge-free
Degree of Supervision	unsupervised	supervised	unsupervised
Corpus required	–	annotated text	raw text

Table 1.1 Characteristics of three paradigms for the computational treatment of natural language

Various combinations of these paradigms lead to hybrid systems, e.g. it is possible to construct the rules needed for CL by statistical methods, or to build a supervised standard NLP system on top of an unsupervised, knowledge-free system, as conducted in Section 6.9. As already discussed earlier, semi-supervised learning is located in between statistical NLP and the SD paradigm.

1.2 Relation to General Linguistics

Since the subject of examination in this work is natural language, it is inevitable to relate the ideas presented here to linguistics. Although in this book, neither one of the dominating linguistic theories is implemented nor something that would be dubbed 'linguistic theory' by linguists is proposed, it is still worthwhile looking at those ideas from linguistics that inspired the methodology of unsupervised natural language processing, namely linguistic structuralism and distributionalism. Further, the framework shall be examined along desiderata for language processing systems, which were formulated by Noam Chomsky.

Serving merely to outline the connection to linguistic science, this section does by no means raise a claim for completeness on this issue. For a more elaborate discussion of linguistic history that paved the way to the SD paradigm, see [42].

1.2.1 Linguistic Structuralism and Distributionalism

Now, the core ideas of linguistic structuralism will be sketched and related to the SD paradigm. Being the father of modern linguistics, de Saussure [211] introduced his negative definition of meaning: the signs of language (think of linguistic elements such as words for the remainder of this discussion) are solely determined by their relations to other signs, and not given by a (positive) enumeration of characterisations, thereby harshly criticising traditional grammarians. This is to say, language signs arrange themselves in a space of meanings and their value is only differentially determined by the value of other signs, which are themselves characterised differentially.

De Saussure distinguishes two kinds of relations between signs: syntagmatic relations that hold between signs in a series in present (e.g. neighbouring words in a sentence), and associative relations for words in a "potential mnemonic series" [211, p. 123]. While syntagmatic relationships can be observed from what de Saussure calls *langage* (which corresponds to a text corpus in terms of this work), all other relationships subsumed under 'associations' are not directly observable and can be individually different. A further important contribution to linguistic structuralism is attributed to Zelling Harris [124, 123], who attempts to discover some of these associative or paradigmatic relations. His *distributional hypothesis* states that words of similar meanings can be observed in similar contexts, or as popularised by J. R. Firth: "You shall know a word by the company it keeps!" [99, p.179][2]. This quote can be understood as the main theme of *distributionalism*: determining the similarity of words by comparing their contexts.

Distributionalism does not look at single occurrences, but rather at a word's distribution, i.e. the entirety of contexts (also: global context) it can occur in. The notion of context is merely defined as language elements related to the word; its size or structure is arbitrary and different notions of context yield different kinds of similarities amongst the words sharing them. The consequences of the study of Miller and Charles [179] allow to operationalise this hypothesis and to define the similarity of two signs as a function over their global contexts: the more contexts two words have in common, the more often they can be exchanged, and the more similar they are. This immediately gives rise to discovery procedures that compare linguistic elements according to their distributions, as conducted in later chapters of this book.

---

[2] Ironically, Firth did not mean to pave the way to a procedure for statistically finding similarities and differences between words. He greatly objected to de Saussure's views and clearly preferred the study of a restricted language system for building a theory, rather than using discovery procedures for real, unrestricted data. In his work, the quote refers to assigning correct meanings to words in habitual collocations.

Grouping sets of mutually similar words into clusters realises an abstraction process, which allows generalisation for both words and contexts via class-based models [cf. 49]. It is this mechanism of abstraction and generalisation, based on contextual clues, that allows the adequate treatment and understanding of previously unseen words, provided their occurrence in well-known contexts.

1.2.2 Adequacy of the Structure Discovery Paradigm

This section aims at providing theoretical justification for bottom-up discovery procedures as employed in the SD paradigm. This is done by discussing them along the *levels of adequacy* for linguistic theories, set up in [59], and examining to what extent this classification applies to the procedures discussed in this work. For this, Chomsky's notions of linguistic theory and grammar have to be briefly sketched. For Chomsky, a (generative) grammar is a formal device that can generate the infinite set of grammatical (but no ungrammatical) sentences of a language. It can be used for deciding the grammaticality of a given sentence [see also 58]. A linguistic theory is the theoretical framework, in which a grammar is specified. Chomsky explicitly states that linguistic theory is only concerned with grammar rules that are identified by introspection, and rejects the use of discovery procedures of linguistic regularities from text corpora, since these are always finite and cannot, in his view, serve as a substitute for the native speaker's intuitions.

Having said this, Chomsky provides a hierarchy of three levels of adequacy to evaluate grammars.

- A grammar with *observational adequacy* accounts for observations by exhaustive enumeration. Such "item-and-arrangement grammars" [59, p. 29] can decide whether a given sentence belongs to the language described by the grammar or not, but does not provide any insights into linguistic theory and the nature of language as such.
- A higher level is reached with *descriptive adequacy*, which is fulfilled by grammars that explain the observations by rules that employ "significant generalizations that express underlying regularities of language" [59, p. 63].
- *Explanatory Adequacy* is the highest level a grammar can reach in this classification. Grammars on this level provide mechanisms to choose the most adequate of competing descriptions, which are equally adequate on the descriptive level. For this, "it aims to provide a principled basis, independent of any particular language" [59, p. 63].

According to Chomsky, the levels of descriptive and explanatory adequacy can only be reached by linguistic theories in his sense, as only theoretic means found by introspection based on the native speaker's intuition can perform the necessary abstractions and meta-abstractions. Criticising exactly this statement is the subject of the remainder of this section.

When restricting oneself to a rule-based description of universal language structure like Chomsky does, there does not seem to be any other option than proceeding in an introspective, highly subjective and principally incomplete way: as already Sapir [209, p. 38] stated: "all grammars leak", admitting the general defect of grammar theories to explain the entirety of language phenomena. Especially when proceeding in a top-down manner, the choice "whether the concept [an operational criterion] delimits is at all close to the one in which we are interested" [59, p. 57] is subjective and never guaranteed to mirror linguistic realities. But when dropping the constraint on rules and attributing explanatory power to bottom-up discovery procedures, the levels of adequacy are also applicable to the SD framework.

Admitting that operational tests are useful for soothing the scientific conscience about linguistic phenomena, Chomsky errs when he states that discovery procedures cannot cope with higher levels of adequacy [59, p. 59]. By using clustering procedures as e.g. described in [49] and in Chapter 4, abstraction and generalisation processes are realised that employ the underlying structure of language and thus serve as algorithmic descriptions of language phenomena. Further, these class-based abstractions allow the correct treatment of previously unobserved sentences, and a system as described in Section 6 would clearly attribute a higher probability (and therefore acceptability) to the sentence "colorless green ideas sleep furiously" than to "furiously sleep ideas green colorless" based on transition probabilities of word classes (examples taken from [59, p. 57]), see also [186] on this issue.

Unlike linguistic theories, the systems equipped with these discovery procedures can be evaluated either directly by inspection or indirectly by measuring their contribution to an application. It is therefore possible to decide for the most adequate, best performing discovery procedure amongst several available ones. Conducting this for various languages, it is even possible to identify to what extent discovery procedures are language independent, and thus to arrive at explanatory power, which is predictive in that sense that explanatory adequate procedures can be successfully used on previously unseen languages.

The great advantage of discovery procedures is that, once defined algorithmically, they produce abstractions that are purely based on the data provided, being more objective and conclusive than rules found by introspection. While simply not fitting in the framework of linguistic theories, they are not phantoms but realities of language, so a complete description of language theory should account for them [see also 1].

In terms of quantity and quality, the goals of linguistic theory and unsupervised discovery procedures are contrary to one another. Linguistic theory aims at accounting for most types of phenomena irrespective of how often these are observed, while application-based optimisation targets at an adequate treatment of the most frequent phenomena. Therefore, in applying unsupervised discovery procedures, one must proceed quantitatively, and not qualitatively, which is conducted in this work at all times.

1.3 Similarity and Homogeneity in Language Data

1.3.1 Levels in Natural Language Processing

Since the very beginning of language studies, it is common ground that language manifests itself by interplay of various levels. The classical level hierarchy in linguistics, where levels are often studied separately, is the distinction of [see e.g. 116] phonological, morphological, syntactic, semantic and pragmatic level. Since this work is only concerned with digitally available written language, levels that have to do with specific aspects of spoken language like phonology are not considered here. Merely considering the technical data format for the text resources used here, basic units and levels of processing in the SD paradigm are [cf. 126]:

- character level
- token level
- sentence level
- document level

The character level is concerned with the alphabet of the language. In case of digitally available text data, the alphabet is defined by the encoding of the data and consists of all valid characters, such as letters, digits and punctuation characters. The (white)space character is a special case, as it is used as delimiter of tokens. Characters as the units of the character level form words by concatenation. The units of the token level are tokens, which roughly correspond to words. Hence, it is not at all trivial what constitutes a token and what does not. The sentence level considers sentences as units, which are concatenations of tokens. Sentences are delimited by sentence separators, which are given by full stop, question mark and exclamation mark. Tokenisation and sentence separation are assumed to be given by a pre-processing step outside of the scope of this work.[3]

A clear definition is available for documents, which are complete texts of whatever content and length, i.e. web pages, books, newspaper articles etc.

When starting to implement data-driven acquisition methods for language data, only these units can be used as input elements for SD processes, since these are the only directly accessible particles that are available in raw text data.

It is possible to introduce intermediate levels: morphemes as subword units, phrases as subsentential units or paragraphs as subdocument units. However, these and other structures have to be found by the SD methodology first, so they can be traced back to the observable units.

Other directly observable entities, such as hyperlinks or formatting levels in web documents, are genre-specific levels that might also be employed by SD algorithms, but are not considered for now.

[3] for experiments in later chapters, the tools of the Leipzig Corpora Collection (LCC, [198]) were used throughout for tokenisation and sentence separation.

1.3.2 Similarity of Language Units

In order to group language units into meaningful sets, they must be compared and similarity scores for pairs of units need to be assigned. Similarity is determined by two kinds of features: *surface features* and *contextual features*. Surface features are obtained by only looking at the unit itself, i.e. tokens are characterised on their surface by the letters and letter combinations (such as character N-grams) they contain, sentences and documents are described with the tokens they consist of. Contextual features are derived by taking the context of the unit into consideration.

The context definition is arbitrary and can consist of units of the same and units of different levels. For example, tokens or character N-grams are contextually characterised by other tokens or character N-grams preceding them, sentences are maybe similar if they occur in the same document, etc. Similarity scores based on both surface and contextual features require exact definition and a method to determine these features, as well as a formula to compute the similarity scores for pairs of language units.

Having computational means at hand to compute similarity scores, language units can eventually be grouped into homogeneous sets.

1.3.3 Homogeneity of Sets of Language Units

While the notion of language unit similarity provides a similarity ranking of related units with respect to a specific unit, a further abstraction mechanism is needed to arrive at classes of units that can be employed for generalisation. Thus, it is clear that a methodology is needed for grouping units into meaningful sets. This is realised by clustering, which will be discussed in depth in Chapter 4. In contrast to the similarity ranking centred on a single unit, a cluster consists of an arbitrary number of units. The advantage is that all cluster members can be subsequently subsumed under the cluster, forming a new entity that can give rise to even more complex entities.

Clustering can be conducted based on the pairwise similarity of units, based on arbitrary features. In order to make such clustering and at the same time generalisation processes successful, the resulting sets of units must exhibit *homogeneity* in some dimensions. Homogeneity here means that the cluster members agree in a certain property, which constitutes the abstract concept of the cluster. Since similarity can be defined along many features, it is to be expected that different dimensions of homogeneity will be found in the data, each one covering only certain aspects, e.g. on syntactic, semantic or morphological levels. Homogeneity is, in principle, a measurable quantity and expresses the plausibility of the grouping. In reality, however, this is very often difficult to quantify, thus different groupings will be judged on their utility in the SD paradigm: on the one hand by their ability to generalise in a way that further structure can be discovered with these abstract concepts forming the building blocks, on the other hand by their utility as features in task-driven applications.

1.4 Vision: The Structure Discovery Machine

The remainder of this chapter is dedicated to an outline of the ultimate goal of SD: a set of algorithmic procedures that encompasses the discovery of the entirety of structure that can be discovered in a data-driven way. Viewing the practical results in Chapters 5, 6 and 7 as being only a starting point, I will now envision a number of capabilities of such a system and discuss them along the usual pipeline of processing steps for language processing. The 'traditional' terms shall serve as an aid for imagination — without doubt, the discovery procedures will not reproduce theoretically pure sub-levels, as indicated above. Nevertheless, linguistic theory can play the role of a source of what is necessary and which phenomena are to be accounted for.

When exhibited to language data, the Structure Discovery Machine (SDM) identifies the basic word and sentence units it will operate on and finds a suitable representation — no matter whether the SDM is exposed to text, speech or other encodings of language. Already here, an abstraction process is involved that groups e.g. different phonetic variants in speech or different character sets in written language. Then, different languages are identified in the stream and the corresponding parts are grouped (see Section 5). In a similar way, domain-specific subsets of the monolingual material are marked as such, e.g. by techniques as employed in document clustering (see [231] for an overview).

For each language, a syntactic model is set up that encompasses parts of speech (cf. Section 6) for basic word units, chunking to phrasal units of several words [as in e.g. 67] and syntactic dependencies between basic and complex units[4] such as in [140; 36; 188, inter al.]. This requires a procedure handling derivation, inflection and compounding of units, realised by a component similar to those compared in the MorphoChallenge [146], or more recently [113]. Contextual analysis of content-bearing units allows to hypothesise different meanings for units with semantically ambiguous use and to disambiguate their occurrences (cf. Section 7). Having circumnavigated the pitfalls of lexical ambiguity, units are classified according to their semantic function and relation. Semantic classes have been previously learned in [e.g. 53; 21; 235]; for semantic relations, [238] shows ways how to extract them from massive corpora. The works of Davidov and Rappoport [77] and Davidov et al. [78] illustrate methods on how to extract sets of semantically similar words and their typical relations from web data. Reoccurring relations between units of similar or different semantic function will be marked as such, serving as something similar to FrameNet [14] annotations. Coreference has been successfully learnt in an unsupervised way by [120], information extraction with self-learned templates is laid out in [56].

In its highest and purest form, the SDM will not even be engineered by humans plugging together discovery procedures based on their intuitions, but will self-assemble by trying out interplays of a parameterisable inventory of procedures, thus optimising the overall structural description in terms of generalisation power. This,

[4] cf. approaches to parsing, such as HPSG [193], LFG [128] and dependency parsing [184].

however, is a much larger project than outlined in this book, and raises several yet unanswered research questions.

Output of the SDM is a multidimensional annotation of the raw input data with labels that denote abstractions of structurally similar phenomena. Some kinds of annotations are orthogonal to each other, others can be arranged hierarchically and many will be dependent on other annotations. The scope of annotations is not limited, ranging from the most basic to the most complex units. This holistic approach to data-driven self-annotation rather overgenerates and is not restricted to a small set of structural phenomena. To determine which generalisations are useful, these have to be tested for utility in task-based evaluations, see next section.

Applications of these annotations are twofold: firstly, they can be employed as features in machine learning for a task-based adaptation: stemming, parts-of-speech tagging, chunking, named entity recognition, information extraction and parsing in their present form will greatly benefit from this rich inventory of features, which will significantly reduce the necessary amount of training instances. Overlapping representations of complex facts give rise to mapping queries to answers in a Question Answering (QA) system (as e.g. conducted in [200]), which is currently a very active area of research. IBM's Watson system, for example, is a system that works massively parallel by simultaneously using many different approaches to question answering [97], and uses machine learning on high-level features to determine its answers. Since names and values of features are irrelevant in this setting besides their ability to grasp the differentiation granularity needed for the task, annotations generated by the SDM can be easily incorporated and evaluated in such an application-based way. Feature selection mechanisms [cf. 144, inter al.] allow to choose those annotations that correlate best with the task-defined structural regularities.

Secondly, when examining what annotations are the most important ones for solving language processing tasks and tracing back their creation to their data-driven origin, one will find that the procedures of the SDM provide insights into the system of constituents of natural language and could play the role of de Saussure's desperately sought realitiesas stated in the epigraph of this chapter.

Elaborating on the example of Question Answering, a complete Structure Discovery version of a QA system is now briefly sketched. The task of a QA system is defined by question-answer pairs that have been constructed manually. The challenge in QA is to connect questions to text passages and extract the answer. This is hampered by the fact that, in most cases, the wording of the questions does not exactly match any of the available text passages. In traditional QA, text passages are preprocessed and indexed in a way that fuzzy matching is supported, e.g. by synonym expansion and grammatical canonicalisation. Further, grammatical patterns and transformations are applied to e.g. transform a question like "Who invented the light bulb?" into patterns like "X invented the light bulb" or "X, the inventor of the light bulb" that can be matched on text passages. The linguistic structures, on which these transformation and matching modules operate, are important in intermediate representations, but do not play a direct role in the answer: it is not important to know that e.g. X is in the subject position of "invented", as long as X can be resolved as the answer to the question. Thus, preprocessing could be replaced by Structure Discovery processes that would add annotations similar to morphology, parts of speech, dependency parses and lexical substitutions to both the text passage and the question. From a multitude of possible realisations and parameterisations of components realising these structural annotations, those are selected that contribute the most in training a system on question answer pairs. For different question types, as e.g. found by clustering on question structure or answer category, subsystems with different components could be learned. Some of the processes will be shared, such as e.g. the morphology component. Others might differ considerably, e.g. domain-specific semantic word classes.

1.5 Connecting Structure Discovery to NLP tasks

Obviously, the iterative building of a stack of Structure Discovery procedures has to be guided by a target application to determine the usefulness of SD annotations. But as opposed to linguistic preprocessing steps, where system behaviour is defined by carefully assigned linguistic annotations by expert annotators, a more task-driven and usage-driven approach to data acquisition is proposed in this section.

The acquisition bottleneck of knowledge processing stems from the fact that it is expensive to create large amounts of high quality data for training or validating automatic systems. There are two ways to widen this bottleneck: either by employing algorithms that better abstract over the data so that less guidance in the form of annotations is needed for similar performance, or by finding a cheaper way of creating the data.

Structure Discovery algorithms perform abstractions and help to minimise the amount of data collection for intermediate processing steps — in the ideal case, the dependency of manual data creation for preprocessing would be removed altogether. For the last step that actually performs the task defined by the application, it might as well be that two sources of data acquisition are suitable, which have not yet been used widely in the NLP community: crowdsourcing and user logs.

Crowdsourcing is a means to distribute work to a potentially large group of workers. On a crowdsourcing platform like Amazon Mechanical Turk[5] or Crowdflower[6], requesters define tasks and farm them out to workers. Compared to professional annotators, these workers are comparatively cheap and produce noisier data sets since quality control is more difficult than with expert annotators. Crowdsourcing recently gained increased attention for creating language processing data sets, see [52] for a survey. Criticism of crowdsourcing linguistic datasets includes, amongst other things, that crowd workers are not linguists and therefore miss subtle distinctions and are not able to produce good enough data for linguistic preprocessing steps.

[5] www.mturk.com [August 31, 2011]

[6] http://crowdflower.com [August 31, 2011]

While it seems not feasible to create e.g. a treebank with crowdsourcing, since workers cannot be sufficiently trained to produce consistent annotations according to a linguistic theory, the picture changes when shifting the preprocessing to Structure Discovery processes and letting untrained workers do what they are good at: be users of a system.

An advanced system of natural language understanding should learn from the users it is operated by — in analogy to today's search engines, that are heavily tuned by mining logs of user behaviour. For example, a spelling correction system could memorise what corrections were accepted by the user in what contexts and use this data for adjusting its model. As this sort of feedback is very indirect and noisy, a large amount of user data is needed to effectively tune such a system. Also, the system needs a certain quality to begin with — users will only operate the system regularly to produce enough usage data if they have a benefit from doing so. In order to reach this quality threshold, the benefit can be provided by paying users via crowdsourcing, which plays the role of a segue between the content-data-driven Structure Discovery layer on the one hand, and user-data-driven systems on the other hand.

1.6 Contents of this Book

This section provides a guideline for reading this book. Depending on whether the reader is more interested in the theoretical underpinnings or the more practical aspects of this work, different parts should be in focus. Chapters 2, 3 and large parts of Chapter 4 are of theoretical nature. These parts are necessary to understand the paths taken in the more application-oriented Chapters 5, 6 and 7, but can probably be merely used for reference if the reader is less interested in constructing her/his own Structure Discovery processes, but rather in using existing SD implementations.

1.6.1 Theoretical Aspects of Structure Discovery

Readers familiar with basic notions of graph theory can safely skip Section 2.1 and merely use it for reference to the exact definitions of notions used in later chapters. Section 2.2 is dedicated to scale-freeness and the Small World property of networks, describing a number of emergent random graph generation models. In Sections 3.1 and 3.2, we look at quantitative characteristics of language data. The notion of the word co-occurrence graph, which reappears frequently in later chapters, is defined in Section 3.2.1. In Section 3.3, an emergent random text model is defined, which models quantitative characteristics of language much more closely than other graph generation models or random text models. This model contributes to the understanding of the origin and the emergence of language structure, but is not directly related to applications described later. Section 4.1 on clustering methods in general and graph clustering methods in particular discusses reasons for the need of a fast graph clustering algorithm, but might not contain a lot of new material for readers familiar with these fields. Understanding the definition of the Chinese Whispers algorithm in Section 4.2, however, will be very helpful for the applied part of the book.

1.6.2 Applications of Structure Discovery

Chapter 5 describes an SD system for language separation. This method is not only able to find small injections of foreign language into monolingual corpora, but also to robustly recognise a large number of languages in highly multilingual data, even for non-standard texts such as Twitter data. The reader is pointed to an implementation of the system. Unsupervised part-of-speech tagging is discussed in high detail in Chapter 6. The utility of using unsupervised word class labels is demonstrated on a number of standard natural language processing tasks. Again, an implementation of the system is available for download, as well as a number of already induced taggers for various languages that can be easily incorporated in NLP systems. The topic of Chapter 7 revolves around word senses and meaning. In this chapter, the task is defined through lexical substitution data collected by crowdsourcing, which is also available for download. Unsupervised word sense induction features improve the performance of a supervised lexical substitution system. Also, the selection process for unsupervised features stemming from different parameterzations of the same SD process is demonstrated.

1.6.3 The Future of Structure Discovery

The final chapter summarises the achievements contained in this volume, and gives an outlook on possible directions of future work in Structure Discovery. With each unsupervised and knowledge-free method added to the inventory of Structure Discovery processes, another step is taken towards a data-driven and application-driven approach to natural language processing, where tight control of the preprocessing pipeline is given up in favour of language-independence and domain-independence.

Chapter 2
Graph Models

Abstract This chapter provides basic definitions of graph theory, which is a well-established field in mathematics, dealing with properties of graphs in their abstract form. Graph models are a way of representing information by encoding it in vertices and edges. In the context of language processing, vertices will denote language units, whereas edges represent relations between these, e.g. a neighbourhood or similarity relation. This way, units and their similarities are naturally and intuitively translated into a graph representation. Note that graph models discussed here are not to be confused with graphical models [133], which are notations that represent random variables as nodes in Bayesian learning approaches. After revisiting notions of graph theory in Section 2.1, the focus is set on large-scale properties of graphs occurring in many complex systems, such as the Small World property and scale-free degree distributions. A variety of random graph generation models exhibiting these properties on their graphs will be discussed in Section 2.2. The study of large-scale characteristics of graphs that arise in Natural Language Processing using graph representations are an essential step towards approaching the data, in which structural regularities shall be found. Structure Discovery processes have to be designed with awareness about these properties. Examining and contrasting the effects of processes that generate graph structures similar to those observed in language data sheds light on the structure of language and their evolution.

2.1 Graph Theory

2.1.1 Notions of Graph Theory

This section contains basic definitions of graph theory that will be necessary in all later chapters. The notation follows [39]. Readers familiar with these concepts can jump to Section 2.2 and use this section for reference.

2.1.1.1 Graph, Vertices, Edges

A graph G is a pair (V,E) of finite sets V and E such that E is a subset of unordered pairs of $V \times V$. V is the set of vertices, E is the set of edges. If G is a graph, then $V = V(G)$ is the vertex set of G, and $E = E(G)$ is the edge set. An edge $\{x,y\}$ is said to join the vertices x and y and is denoted by xy, also $e(x,y)$. Thus, xy and yx mean exactly the same edge, x and y are called endpoints of that edge, x and y are said to be adjacent vertices. Equivalently, it is said that an edge connects x and y, or e is incident with x and y. In the following, the graphs are restricted to being simple, that is no self-loops (irreflexive) and no parallel edges, unless explicitly stated otherwise.

2.1.1.2 Subgraph

A subgraph $G(S)$ of a graph G is induced by a set of vertices $S \subset V(G)$ and contains all edges of G that connect $s_i, s_j \in S$. The set of edges of $G(S)$ is denoted by $E(S)$.

2.1.1.3 Order and Size

The order of G is the number of vertices in G, denoted by $|G|$ or n, also written $|V(G)|$. Here, $|.|$ is the size of a set. The size of G is the number of edges, denoted by $|E(G)|$ or m. We write G^n for a graph of order n. $G(n,m)$ denotes a graph of order n and size m.

2.1.1.4 Subsets of Vertices

Given two disjoint subsets U and W of the vertex set of a graph $V(G)$, the set of edges joining vertices from U and W is written $E(U,W)$. The number of edges in $E(U,W)$ is $|E(U,W)|$.

2.1.1.5 Neighbourhood of a Vertex

The neighbourhood of a vertex $v \in V(G)$ is the induced subgraph of the vertices adjacent to v, that is $neigh(v) = \{x|\{v,x\} \in E(G)\}$. Sometimes this is called open neighbourhood in contrast to the closed neighbourhood formed by $neigh(v) \cup \{v\}$.

2.1.1.6 Path

A path is a graph P of the form $V(P) = \{x_0, x_1, ..., x_l\}$, $E(P) = \{x_0x_1, x_1x_2, ..., x_{l-1}x_l\}$. The path is denoted by $x_0x_1..x_l.$, l is the length of the path P, which is the number of edges it contains. P is said to be a path from x_0 to x_l or an $x_0 - x_l$-path.

2.1.1.7 Distance

The distance $d(x,y)$ between vertices x and y is the length of a shortest $x-y$ path. If no such path exists, $d(x,y)=\infty$. The distance between a vertex and itself is $d(x,x)=0$.

2.1.1.8 Connectivity and Components

A graph is connected, if for every pair $\{x,y\}$ of distinct vertices $x \in V(G)$, $y \in V(G)$, $x \neq y$ there exists a path from x to y. A maximal connected subgraph of G is called a component of G. A cutvertex is a vertex whose deletion increases the number of components. An edge is called bridge, if its deletion increases the number of components.

2.1.1.9 Partitions

A partition of a graph $G(V,E)$ is a set of disjoint sets of vertices $\{P_1,P_2,...P_n\}$ with $P_i \subset V$ such that for all $i,j \in \{1..n\}$, $i \neq j$: $P_i \cap P_j = \emptyset$ and $\bigcup_{i=1..n} P_i = V$. The sets P_i are called parts, also clusters. These two terms are used interchangeably in the remainder of this work.

2.1.1.10 Cuts

Any subset of vertices $S \subset V$, $S \neq \emptyset$ creates a cut, which is a partition of V into two disjoint subsets S and $V \setminus S$. The size of a cut S of graph G is defined as $c_G(S) = e(S,V \setminus S)$. It measures the number of edges that have to be eliminated in order to obtain the two components S and $V \setminus S$ from G.

2.1.1.11 Bipartite Graph

A graph G is bipartite with vertex subsets V_1 and V_2 if $V(G) = V_1 \cup V_2, V_1 \cap V_2 = \emptyset$ and every edge joins a vertex from V_1 and a vertex from V_2. Similarly, G is r-partite with vertex classes $V_1,V_2,...V_r$, if $V(G) = \bigcup_{1..r} V_i$ and $V_i \cap V_j = \emptyset$ for all $i,j \in \{1..n\}$, $i \neq j$, and no edge joins vertices of the same vertex part.

2.1.1.12 Complete Graph

It all vertices are pairwise adjacent, then a graph is complete. In particular, a complete graph with n edges has $\binom{n}{2}$ edges (recall that self-loops and parallel edges are excluded), and is denoted by K^n. K^n is also called *n-clique*.

2.1.1.13 Directed Graph

If the edges are ordered pairs of vertices (x,y) instead of unordered pairs $\{x,y\}$, the graph is called directed. An ordered pair (x,y) is said to be a directed edge from x to y, or beginning at x and ending at y, denoted by xy.

2.1.1.14 Edge and Vertex Weight

A graph G is called edge-weighted if there exists a function $ew : E(G) \rightarrow \mathbb{R}$ that assigns an edge weight w_{xy} to every edge $\{x,y\} \in E(G)$. $\{x,y\}$ is said to have the weight $ew(x,y) = w_{xy}$. In analogy, G is vertex-weighted, if there is a function $vw : V(G) \rightarrow \mathbb{R}^+$ that assigns a vertex weight $vw(v)$ to each vertex $v \in V(G)$. Unweighted graphs are a special case of weighted graphs with all vertex and edge weights set to 1.

2.1.1.15 Size of Cut in Weighted Graphs

The size of a cut $c_Gw(S)$ induced by a subset of vertices $S \subset V$ in a weighted graph is defined as the sum of edge weights between S and $V \setminus S$: $c_Gw(S) = \sum_{a \in S} \sum_{b \in V \setminus S} ew(a,b)$. In case of $ew(a,b) \notin E$, $ew(a,b) = 0$.

2.1.1.16 Edge and Vertex Type

A graph G is edge-typed, if there exists a function $et : E(G) \rightarrow S$ with S set of edge types that assigns a type to every edge. A graph is vertex-typed, if there is a function $vt : V(G) \rightarrow T$ with T set of vertex types, such that every vertex $v \in V(G)$ is assigned a type from T. The entirety of vertex type assignments is called vertex-typisation: A vertex typisation of G induces a partition of $G : V(G) = V_1 \cup .. \cup V_n, V_i \cap V_j = \emptyset$ for all $i, j \in \{1..n\}$, $i \neq j$, and for all $x \in V(G), y \in V(G)$: if $x \in V_i$ and $y \in V_i$, then $vt(x) = vt(y)$.

2.1.1.17 Adjacency Matrix

The adjacency matrix of a graph G is a $n \times n$ matrix A_G associated with G where $(a_{ij}) = 1$ if there exists an edge between vertices v_i and v_j, $(a_{ij}) = 0$ otherwise. For edge-weighted graphs, $(a_{ij}) = ew(i, j)$, and $(a_{ij}) = 0$ if no edge exists.

2.1.1.18 Geometrical Representation

A more intuitive way of specifying a graph is to depict it in a geometrical representation. Figure 2.1 illustrates various concepts of graph theory in different notations. Notice that a graphical representation defines a graph unambiguously, but there are many possible graphical representations that define the same graph.

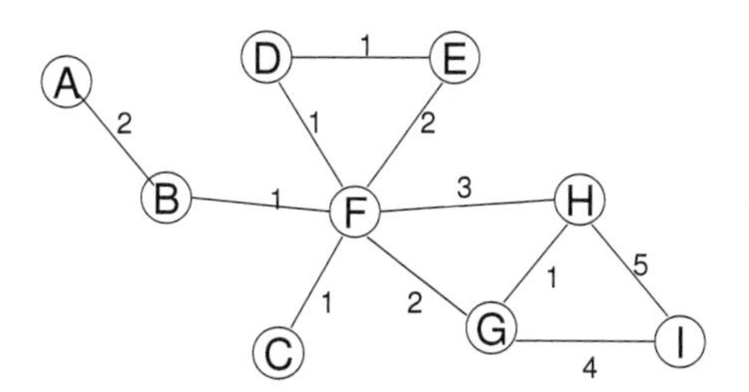

Geometrical representation

a_{ij}	*A*	*B*	*C*	*D*	*E*	*F*	*G*	*H*	*I*
A	0	2	0	0	0	0	0	0	0
B	2	0	0	0	0	1	0	0	0
C	0	0	0	0	0	1	0	0	0
D	0	0	0	0	1	1	0	0	0
E	0	0	0	1	0	2	0	0	0
F	0	1	1	1	2	0	2	3	0
G	0	0	0	0	0	2	0	1	4
H	0	0	0	0	0	3	1	0	5
I	0	0	0	0	0	0	4	5	0

Adjacency matrix

G(V,E)

V={A,B,C,D,E,F,G,H,I}

E={AB,BF,CF,DE,DF,EF,FG,FH,GH,GI,HI}

ew(BF)=ew(CF)=ew(DE)=ew(DF)=ew(GH)=1;

ew(AB)=ew(EF)=ew(FG)=2; ew(FH)=3;

ew(GI)=4; ew(HI)=5;

Set representation

Fig. 2.1 Example for representing a graph in geometrical, adjacency matrix and set representation. B and F are cutvertices, AB, BF and CF are bridges. Another possibility is the representation in adjacency lists

2.1.2 Measures on Graphs

A variety of measures exist that characterise certain properties of graphs. This overview here is by no means complete and merely defines measures that are needed at a later stage. For an exhaustive treatment of general graph measures, please refer to [167, inter al.], and for a survey on measures used for complex networks, see [70].

Measures on graphs can be divided into local and global measures. Local measures characterise single vertices. In global measures, characteristics of the graph as a whole are combined into a single coefficient or into a distribution.

2.1.2.1 Degree, In-degree, Out-degree

The number of edges incident to v, called degree of vertex v, is $k(v) = |neigh(G)|$. A vertex of degree 0 is called isolated, also singleton. For vertices x of directed graphs, the in-degree $k_{in}(x)$ is defined as the number of edges ending at x, the out-degree $k_{out}(x)$ is the number of edges starting with x.

The average vertex degree $< k >$ of graph $G(V,E)$ is obtained by averaging over the degrees of all vertices: $< k >= \frac{\sum_{v \in V} k(v)}{|V|}$. Notice that a distinction between average in-degree and average out-degree is superfluous.

2.1.2.2 Shortest Path Length and Diameter

The shortest path length between two vertices v and w is defined as the distance $d(v,w)$ between v and w. An efficient way to compute the shortest path length is the *Dijkstra-algorithm*.

The average shortest path length L of a graph $G(V,E)$ is given by averaging the shortest path lengths over all vertex pairs in the same component. For components $V_1,...,V_n$, it is defined as:

$$L = \frac{\sum_{v \in V} \sum_{w \in V, d(v,w) < \infty} d(v,w)}{\sum_{i=1..n} |V_i| \times (|V_i| - 1)} \tag{2.1}$$

The diameter D of a graph $G(V,E)$ is defined as the length of the longest of all shortest paths of G: $D = \max_{v \in V, w \in V, d(v,w) < \infty} d(v,w)$.

2.1.2.3 Clustering Coefficient

Introduced by Watts and Strogatz [244], the vertex clustering coefficient c for a vertex v with $k(v) \geq 2$ in graph G measures the connectivity of vertices in the neighbourhood of v:

$$c(v) = \frac{2 \times |\{\{u,w\} \in E(G), u \in neigh(v), w \in neigh(v)\}|}{|neigh(v)| \times (|neigh(v)| - 1)}. \tag{2.2}$$

The clustering coefficient $C(G)$ of graph G is defined as the average vertex clustering coefficient:

$$C(G) = \frac{1}{|V'|} \sum_{v \in V, k(v) \geq 2} c(v) \tag{2.3}$$

where V' is the set of vertices with degree ≥ 2. Vertices with degree 1 or 0 are not taken into account.

2.1.2.4 Transitivity

The transitivity $T(G)$ of a graph G as defined by Newman et al. [182] is the number of fully connected triplets of vertices divided by the total number of vertex triplets:

$$T(G) = \frac{\sum_{v \in V} \delta(v)}{\sum_{v \in V} \binom{k(v)}{2}} \tag{2.4}$$

with $\delta(v) = |\{\{u,w\} \in E \text{ and } \{v,u\} \in E \text{ and } \{v,w\} \in E\}|$.

2.1.2.5 Relation between Clustering Coefficient and Transitivity

Despite other claims, transitivity is not equivalent to the clustering coefficient, only for graphs where all vertices have the same degree or the same local clustering coefficient, see [213]. This originates from the fact that $C(G)$ is computed by averaging over the vertex clustering coefficients, with each vertex contributing equally to $C(G)$. For $T(G)$, vertices with higher degrees take part in more vertex triplets and thus contribute more in relative terms to $T(G)$. In [213], a weighted clustering coefficient that equals transitivity is proposed: here, vertices v are weighted by the number of "possible opposite edges", that is $k(v)(k(v)-1)/2$. The study further discusses cases of graphs where $C(G)$ and $T(G)$ differ considerably. For the random graph models as discussed in Section 2.2, the most interesting case is a high value for $C(G)$ coupled with a low value of $T(G)$. Figure 2.2 exemplifies a graph where $C \to 1$ and $T \to 0$ as $n \to \infty$. Less extreme cases for high $C(G)$ and low $T(G)$ are characterised by low vertex clustering coefficients for vertices with high degrees.

For determining $C(G)$ and $T(G)$, directed graphs are transformed into undirected graphs. Throughout this work, the efficient implementation as described in [214] is used for computation of these measures.

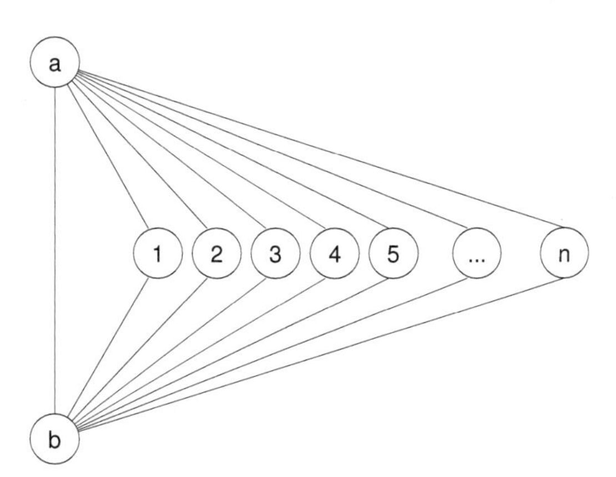

Fig. 2.2 Example of a graph with $C \to 1$ and $T \to 0$ as $n \to \infty$. See also [213]

2.1.2.6 Degree Distribution

The degree distribution $P(k)$ of graph $G(V,E)$ is a probability distribution that provides the probability of choosing a vertex with degree k when randomly selecting a vertex from V with uniform probability. The in-degree and out-degree distributions are defined in analogy. When plotting the degree distribution as in Section 2.2, the number of vertices per degree is depicted. This can be transformed into the degree distribution by normalising with $|V|$.

2.1.2.7 Component Size Distribution

The component size distribution $Comp(x)$ of graph $G(V,E)$ is a probability distribution that provides the probability of choosing a component of size x when randomly selecting a component from V with uniform probability. When plotting the component size distribution, the number of components per size is depicted. This can be transformed into the component size distribution by normalisation.

Table 2.1 summarises the symbols used throughout this chapter, largely following the notation of Steyvers and Tenenbaum [232].

Table 2.1 Symbols referring to graph measures used in this work

Symbol	Explanation
n	number of vertices
L	average shortest path length
D	diameter of the graph
C	clustering coefficient
T	transitivity
k, k_{in}, k_{out}	the degree, the in-degree, and out-degree
$P(k), P(k_{in}), P(k_{out})$	degree distributions
$<k>$	average degree

The graph in Figure 2.1 has the following values for the characteristics introduced: $n = 9$, $L = 2$, $D = 4$, $C = 0.638095$, $T = 0.45, <k> = \frac{22}{9}$ and the degree distribution is given in Table 2.2.

Table 2.2 Degree distribution for graph in Figure 2.1

Degree k	1	2	3	4	5	6
Vertices per degree	2	4	2	0	0	1
$P(k)$	$\frac{2}{9}$	$\frac{4}{9}$	$\frac{2}{9}$	0	0	$\frac{1}{9}$

2.2 Random Graphs and Small World Graphs

The previous section provided definitions and measures that characterise graphs. Now we turn to the question of how graphs can be generated and how properties of the generation process influence the characteristics of the resulting graphs. Here, generation processes are emergent: the graphs are constructed from scratch, without taking data from real-world networks into account. Emergent processes that reproduce characteristics of real-world data, without being informed by the data, possess explanatory power regarding the mechanisms that form the real-world data.

As graphs can be perceived as abstractions of entities and their relations, a generative process is concerned with describing how new entities and relations are added to existing graphs in a systematic way. Different generation principles lead to graphs with different characteristics, and various graph generation models have been proposed in order to yield graphs that resemble observations on real-world data. Now, the most prominent graph generation models in the literature are reviewed. The connection to language as the matter of interest will be indicated, but carried out more thoroughly in Chapter 3.

2.2.1 Random Graphs: Erdős-Rényi Model

The earliest and most basic model of random graph generation in graph theory is the Erdős-Rényi-model (ER-model, Erdős and Rényi [88]). The generation process starts with n vertices and no edges. Each of the possible number of $\frac{n(n-1)}{2}$ undirected vertex pairs gets connected by an edge with probability p. The higher p, the denser gets the graph, with the vertex degrees forming a binomial distribution around their mean $<k>= p(n-1)$.

For comparing graph models, the measures as defined in Section 2.1.2 are employed. Figure 2.3 shows the characteristics of two ER random graphs with 10,000 vertices and $p = 0.001$ ($p = 0.005$). Not surprisingly, the values for transitivity and clustering coefficient hardly differ, as most values for vertex degrees $k(v)$ are close to their mean and the vertex clustering coefficient is directly dependent on p and not on $k(v)$.

The ER model triggered the theory of random graphs. Virtually all studies in this field before the mid-nineties of the 20th century are based on this model, see [38] for an extensive overview. As shall be clear soon, however, the ER model does not capture the characteristics of many graph abstractions of natural phenomena. Another class of graphs that comes closer to what is observed in natural and artificial networks are Small World graphs (SWGs), which will be discussed throughout the remainder of this section.

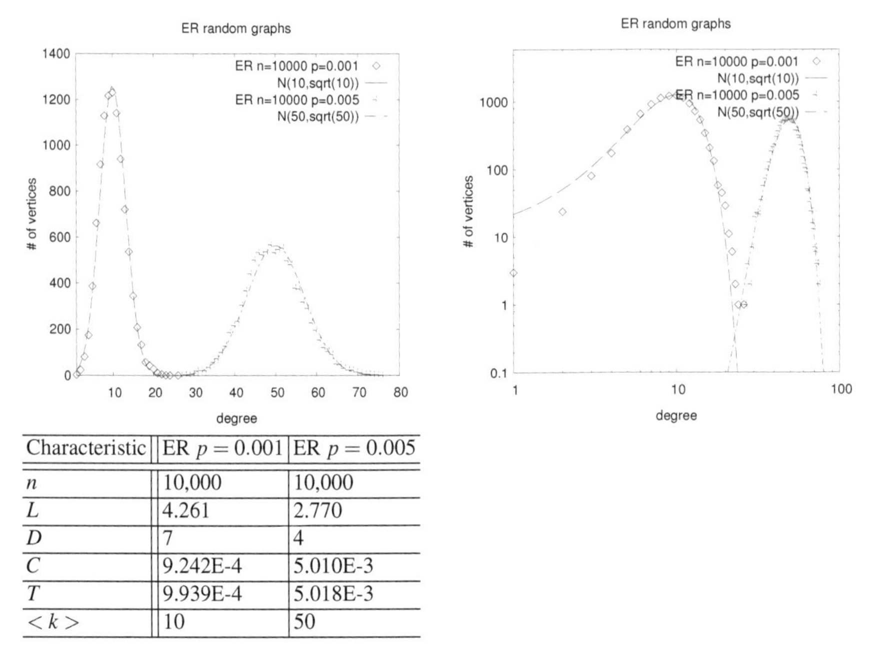

Characteristic	ER $p = 0.001$	ER $p = 0.005$
n	10,000	10,000
L	4.261	2.770
D	7	4
C	9.242E-4	5.010E-3
T	9.939E-4	5.018E-3
$<k>$	10	50

Fig. 2.3 Characteristics of two ER random graphs. The degree distributions are given in a linear scale and in a log-log plot and follow a binomial distribution, which is approximated by a normal distribution $N(<k>, \sqrt{<k>})$ in the plots

2.2.2 Small World Graphs: Watts-Strogatz Model

The first report on Small World phenomena was released by Milgram [176], who performed his well-known experiment using the social network graph between people: addressing 60 letters to a stockbroker living in Boston, he instructed various recruits in Omaha, Nebraska, to merely hand the letter to an acquaintance they would think who could most likely know the receiver. Surprisingly, after six transfers in average, a part of the letters had reached their destination, which was popularised as "six degrees of separation": anyone knows all people on earth over six connections (at least within the USA). In subsequent years, this experiment was repeated under various conditions, confirming this claim. Observations on social networks (as e.g. the entirety of acquaintances in Milgram's experiment) showed, that their corresponding graphs are characterised by a high clustering coefficient C while retaining similar or slightly larger values of L as compared to a Erdős-Rényi random graph with similar numbers of vertices and edges.

This could only be obtained in the Erdős-Rényi model by a probability p that would render us all 'famous party animals': low values of L in hand with high values of C can only be realised using high p values in graph generation, leading to a large $<k>$.

To account for this discrepancy, SWGs were first defined by Watts and Strogatz [244], where a number of interesting graphs are described as having the following property: $n \gg k \gg ln(n) \gg 1$, where $k \gg ln(n)$ guarantees that the graph is connected. They provide a rewiring procedure that constructs a SWG by rewiring edges of a ring lattice with all vertices having the same degree with a certain probability to randomly chosen vertices. When incrementing p, the characteristic path length L drops dramatically, since shortcuts are introduced in the network. The high clustering coefficient C in the regular ring lattice graph drops very slowly with higher p, resulting in the desired characteristics for SWGs.

Since their discovery, graphs with Small World structure have been observed in graphs corresponding to data as diverse as the topology of food webs, electrical power grids, cellular and metabolic networks, the World Wide Web, the internet backbone, neural networks, telephone call graphs, co-authorship and citation in networks of scientists, movie actor networks and overlapping boards of large companies, see [233] for reference. The omnipresence of SWGs in networks whose growth is not governed by a centralised organisation but emerge 'naturally' indicates that understanding their formation processes unveils general principles of nature.

The resulting SWGs are proposed as more accurate models for a large number of self-organising systems, as clustering coefficients are higher while retaining similar L values (cf. Figure 2.3). Figure 2.4 shows characteristics of WS-graphs. Again, transitivity and clustering coefficient highly agree for the same reasons as for the ER model.

2.2.3 *Preferential Attachment: Barabási-Albert Model*

Starting from the observation that in many self-organising systems the degree distribution decays in a power-law, following $P(k) \sim k^{-\gamma}$ (γ = slope of the degree distribution when plotted on log-log scale), Barabási and Albert [16] introduce their graph growing model. Unlike in the ER and WS models, where the number of vertices is fixed in advance, the Barabási-Albert (BA) model starts with a small number of vertices and no edge and iteratively introduces new vertices to the graph, connecting them to a fixed number of existing vertices with a probability based on the degree of the existing vertices. Increasing the probability of connection to 'popular' vertices (i.e. vertices with high degrees) is called *preferential attachment*. Barabási and Albert [16] show that preferential attachment makes the difference between an exponentially decreasing degree distribution and a power-law distribution with $\gamma = 3$ for the standard BA-model as proven in [37]. Further, they give indications how to obtain scale-free SWGs with different power-law exponents. Figure 2.5 shows the characteristics of BA-modelled graphs. In graphs generated by the BA model, the values for transitivity and clustering coefficient are about the same, yet very low as compared to the WS model.

Graphs with a power-law degree distribution are called scale-free for the following reason: in scale-free graphs, there is a significant number of vertices with very

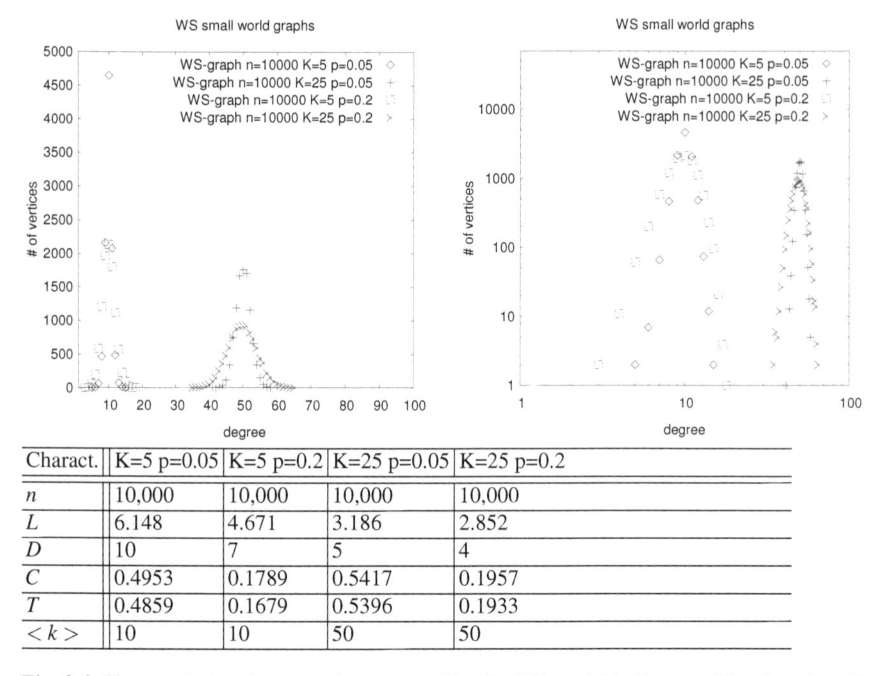

Charact.	K=5 p=0.05	K=5 p=0.2	K=25 p=0.05	K=25 p=0.2
n	10,000	10,000	10,000	10,000
L	6.148	4.671	3.186	2.852
D	10	7	5	4
C	0.4953	0.1789	0.5417	0.1957
T	0.4859	0.1679	0.5396	0.1933
$<k>$	10	10	50	50

Fig. 2.4 Characteristics of two graphs generated by the WS model in linear and log-log plot. The degree distribution decreases exponentially with k, as in the ER-model

high degrees, called *hubs*, whereas in SWGs with an exponential tail of $P(k)$, no such hubs are observed. The scale-free-ness manifests itself in the non-existence of a characteristic vertex degree, i.e. all degrees are present at the same strength, the number of edges for groups of vertices of similar degree is approximately equal. Formally, the degree distribution $P(k) = Ak^{-\gamma}$ remains unchanged to within a multiplying factor when k is multiplied with a scaling factor, i.e. $P(ax) = bP(x)$. This can be seen in Figure 2.5, where the characteristics of graphs generated by the BA-model are depicted. The exact degree distribution is given on the left side of the figure. To obtain a smoother plot, the x-axis was divided into intervals exponentially increasing in size, and the fraction of vertices per degree is given in the figure on the right. For the remainder, this process called *logarithmic binning* is carried out for most figures depicting degree distributions.

2.2.4 Ageing: Power-laws with Exponential Tails

Amaral et al. [11] observe three kinds of natural Small World networks: scale-free graphs as characterised by a power-law distribution of vertex degrees, broad-scale graphs as power-law distributed vertex connectivity with a sharp cut-off, and single-

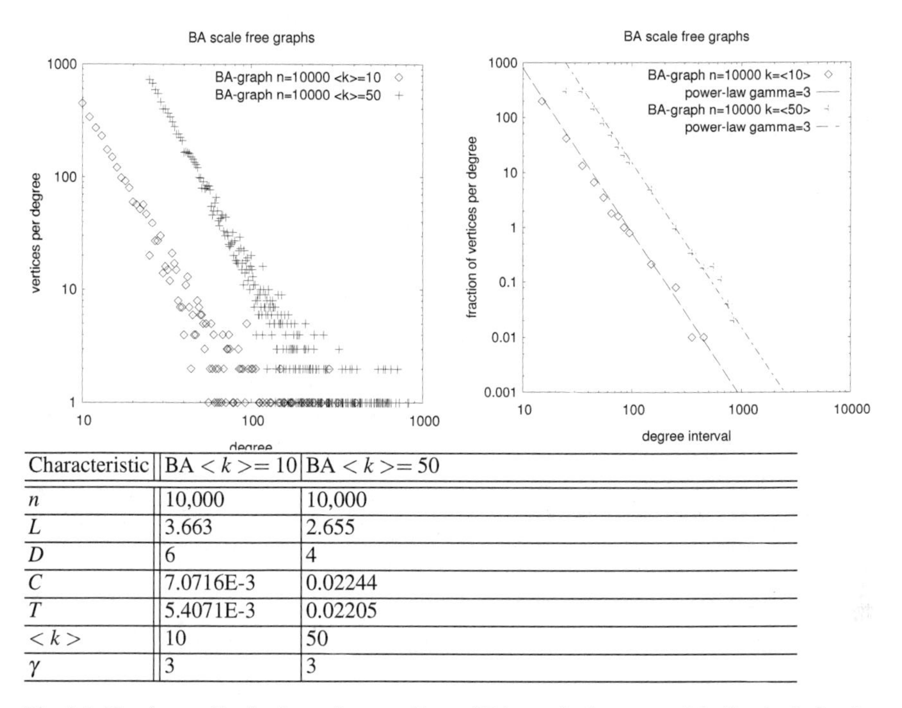

Characteristic	BA $<k>= 10$	BA $<k>= 50$
n	10,000	10,000
L	3.663	2.655
D	6	4
C	7.0716E-3	0.02244
T	5.4071E-3	0.02205
$<k>$	10	50
γ	3	3

Fig. 2.5 The degree distributions of two undirected BA-graphs form a straight line in the log-log plots, indicating that $P(k)$ follows a power-law distribution $P(k) \sim k^{-3}$. Notice that the clustering coefficients are higher than in an ER model graph of the same size and order, but lower than in the graphs generated by the WS-model

scale graphs with faster decaying tails of the degree distribution. These three different varieties of SWGs can be produced by introducing a limiting factor when adding new edges. In the BA-model, an early vertex can obtain an unlimited number of edges as the generation process goes along. In the real world that we seek to model, however, this might be impossible: whereas in a citation graph, a seminal paper can be cited by many following works without any problem, the capability of people to know a large number of acquaintances is limited naturally by their lifetime. Dependent on the limit, the scale-free property might hold for a certain range of connectivity. These findings are exemplified on the movie-actor network (broad-scale as the number of movies per actor is limited yet can reach a high number) and on high school friendships (single-scale because of a smaller number of participants). The authors provide a methodology of generating graphs of all three varieties by introducing the concept of ageing for vertices or alternatively adding cost to connections to vertices proportional to the vertex degree into the BA-model.

2.2.5 Semantic Networks: Steyvers-Tenenbaum Model

Exploring the structure of semantic networks, which now links this section to language data, is the starting point of Steyvers and Tenenbaum [232]. The authors analyse the graph structures of a human association network, Roget's Thesaurus [204] and WordNet [178], finding that all three resources exhibit the characteristics of a scale-free SWG. As they observe much higher clustering coefficients C than predicted by the BA-model, they propose their own network growing algorithm, henceforth called ST-model. It generates scale-free SW graphs in the following way: we start with a small number of fully connected vertices. When adding a new vertex, an existing vertex u is chosen with a probability proportional to its degree. The new vertex is connected to M vertices in the neighbourhood of u. The generative model is parameterised by the number of vertices n and the network's mean connectivity, which approaches $2M$ for large n. Steyvers and Tenenbaum propose a directed and an undirected variant of this model and show high agreement with the characteristics of the semantic networks they examine. The directed variant is obtained from the undirected one by creating a directed edge from the new vertex with a probability d and to the new vertex with a probability $(1-d)$.

The main difference between the ST-model and the BA-model is that the ST-model enforces a high clustering coefficient by attaching all connections of a new vertex in the same neighbourhood. Figure 2.6 shows characteristics of graphs generated by the ST model.

In graphs generated by the ST-model, the transitivity values were measured at about one third of the clustering coefficient of the same graph, which indicates that the vertex clustering coefficients for vertices with high degrees are lower than those of low degrees. This follows from the construction principle: vertices in the neighbourhood of high degree vertices, which have themselves most probably a high degree, get linked to many new vertices, which are themselves interlinked scarcely. Therefore, the vertex clustering coefficient of the high degree vertices is lower than for newer, low-degree vertices, which mostly connect to highly interlinked old vertices.

2.2.6 Changing the Power-Law's Slope: (α,β) Model

The BA-model and the ST-model both produce power-law degree distributions with a slope of $\gamma = 3$, which is an invariant of their models rather than of SWGs in nature (see e.g. examples in Section 3.2 for language data). A more flexible model, called the (α,β)-model, is defined by Kumar et al. [145]. In this growth model for directed SWGs, α is a parameter for specifying the slope of the in-degree distribution $\gamma_{in} = 1/(1-\alpha)$, β is its analogue for the out-degree distribution $\gamma_{out} = 1/(1-\beta)$, with $\alpha,\beta \in [0,1)$. However, slopes of $\gamma < 2$ cannot be generated, as determined empirically and proven for all preferential linking methods in [84]. A further contribution of Dorogovtsev and Mendes [84] is to explain power-law slope variation by

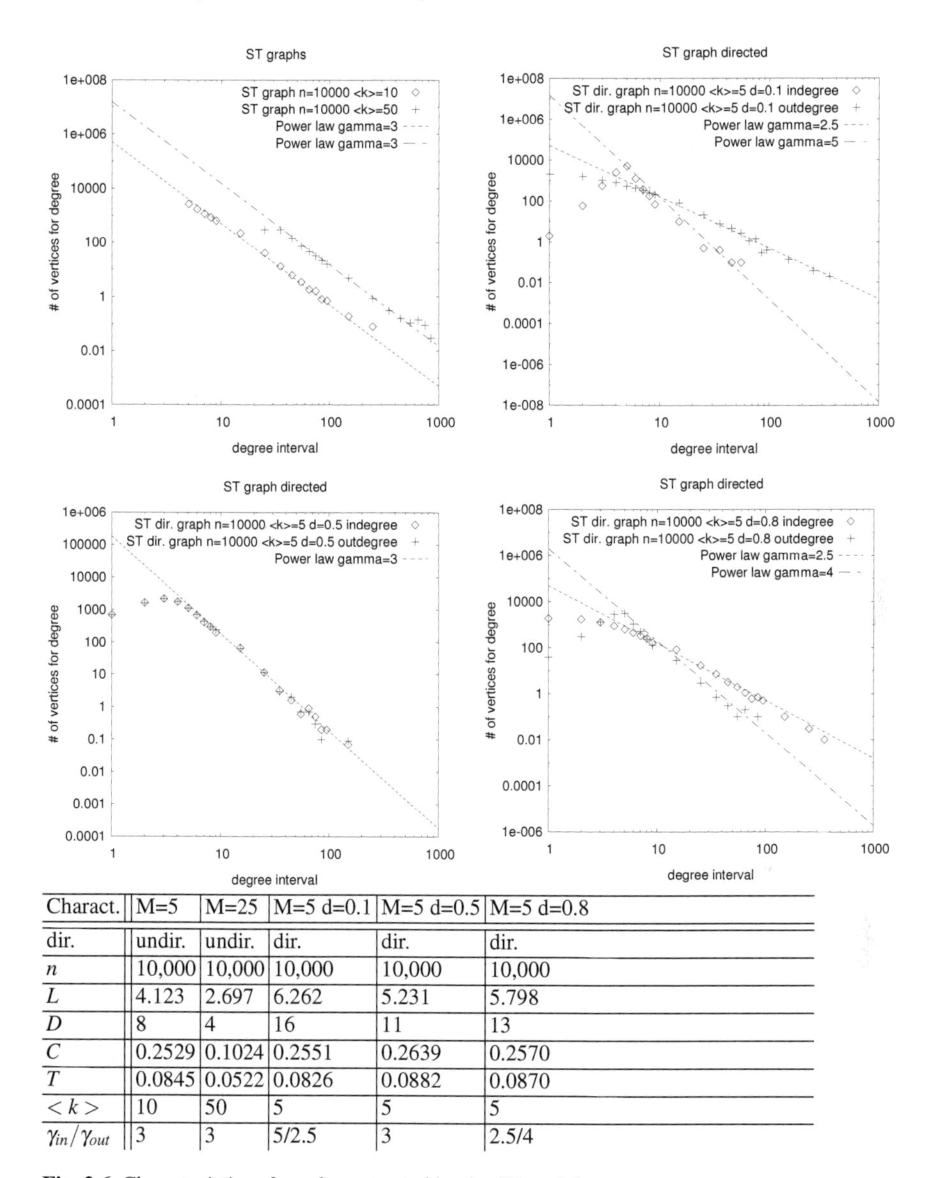

Charact.	M=5	M=25	M=5 d=0.1	M=5 d=0.5	M=5 d=0.8
dir.	undir.	undir.	dir.	dir.	dir.
n	10,000	10,000	10,000	10,000	10,000
L	4.123	2.697	6.262	5.231	5.798
D	8	4	16	11	13
C	0.2529	0.1024	0.2551	0.2639	0.2570
T	0.0845	0.0522	0.0826	0.0882	0.0870
$<k>$	10	50	5	5	5
γ_{in}/γ_{out}	3	3	5/2.5	3	2.5/4

Fig. 2.6 Characteristics of graphs generated by the ST model.

a comparison between growth in the number of vertices and growth in the number of edges: if the total number of edges increases faster than the number of vertices — at this increasing the average degree — the exponent of the degree distribution deviates from $\gamma = 3$.

The (α, β)-model is motivated by the observation that the web graph contains a high number of bipartite cores, which are sets of pages that can be separated into two groups with many links between these groups. These structures emerge in the following building process: each time a new vertex v is created, a fixed number of edges is added following this method: two random numbers $r_1, r_2 \in [0,1]$ are drawn. If r_1 falls within the interval $[0,\alpha]$, the destination of the edge is the new vertex v, otherwise the destination is the destination of a randomly chosen edge. The source of the edge is determined using r_2: if r_2 is in $[0,\beta]$, then the source is v, otherwise it is the source of a randomly chosen edge.

Notice that randomly choosing edges in case of $r_1 > \alpha$ ($r_2 > \beta$) realises preferential attachment, as vertices with higher in-degree (out-degree) more likely become the destination (source) of the new edge. With the possibility to tailor the degree distributions directly to arbitrary slopes in $[2,\infty)$, the (α,β)-model captures the degree distributions of the web graph. This is the only graph generation model discussed here that is reflexive, i.e. edges are allowed to have the same vertex as source and destination. When measuring characteristics, these edges are ignored. The average degree $<k>$ is dependent on the values of α and β. Figure 2.7 shows characteristics for graphs with different α and β values. As a consequence of the construction process, a constant fraction of vertices with both $k_{in} = 0$ and $k_{out} = 0$ can be found in graphs generated by the (α,β)-model. For low values of either α or β, the power-law's slope for the undirected version of the graph is steep. With no mechanism in the construction process that connects vertices adjacent to vertices with high degrees, the discrepancy between C and T is very large.

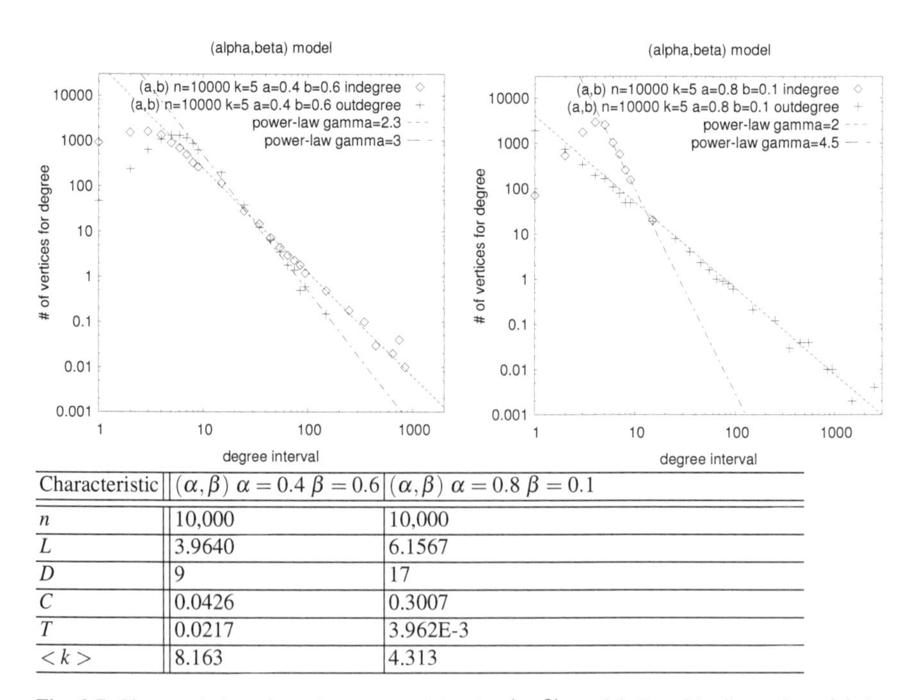

Characteristic	(α,β) $\alpha = 0.4$ $\beta = 0.6$	(α,β) $\alpha = 0.8$ $\beta = 0.1$
n	10,000	10,000
L	3.9640	6.1567
D	9	17
C	0.0426	0.3007
T	0.0217	3.962E-3
$<k>$	8.163	4.313

Fig. 2.7 Characteristics of graphs generated by the (α,β)-model. For this directed model, in-degree and out-degree distributions are given separately.

2.2.7 Two Regimes: Dorogovtsev-Mendes Model

In co-occurrence networks of natural language, sometimes a degree distribution that follows power-laws with two different exponents is observed. The Dorogovtsev-Mendes (DM) model [83] captures this effect in a generation process for undirected graphs. A DM-generated graph is built in the following way: at each time step t, a new vertex is introduced and connected to one present vertex by preferential attachment, i.e. with a probability proportional to the old vertex' degree. This step is equal to the BA-model with $< k >= 2$. But in difference to the BA-model, ct edges between old vertices are introduced in every time step t between previously not connected old vertices i and j, with a probability according to the product of their vertex degrees $k(i) \times k(j)$.

It has been shown in [83] that the average degree is dependent on the time step and is given by $< k >= 2 + ct$. For $1 \ll ct$, the degree distribution is governed by two power-laws with $\gamma_1 = 1.5$ for low degrees and $\gamma_2 = 3$ for high degrees, with a crossover point at $k_{cross} =\approx \sqrt{ct}(2+ct)^{1.5}$. This is achieved by mixing two different growth rates of edges as compared to vertices [see 84]: a constant growth for edges involving the newly introduced vertices and an increasing growth rate for edges between old vertices. Dorogovtsev and Mendes successfully model the word web as described by Ferrer-i-Cancho and Solé [95]: word tokens represent vertices and edges are introduced if their corresponding endpoints are found in adjacent positions in the underlying text corpus. These *word co-occurrence graphs* will be subject to deeper exploration in Section 3.2.

Figure 2.8 displays the degree distribution and the characteristics of graphs generated by the DM-model. The DM-model generates scale-free Small World graphs with two power-law regimes. Values for C and T are much higher than in a BA-model graph. As opposed to the ST model, a higher $< k >$ in the DM-model leads to higher C and T values. Throughout, transitivity T is measured somewhat lower than the clustering coefficient C. Choudhury et al. [60] take this analysis further by examining the topology of co-occurrence networks beyond the two-power-law regime and elaborating on a core-periphery structure.

2.2.8 Further Remarks on Small World Graph Models

In [8], no generation model is provided, but statements about the component size distribution of scale-free graphs for different degree distribution slopes γ are proven. While for $\gamma < 1$, the graph is almost surely fully connected, a γ between 1 and 2 produces one very large component and some small components of constant size. The range of $2 < \gamma < 3.4785$[1] delimits a family of graphs with small components in a size order of the logarithm of the number of vertices, and $\gamma > 3.4785$ yields graphs with component size distributions following a power-law as well.

[1] numerical result of a complex expression as given in [8].

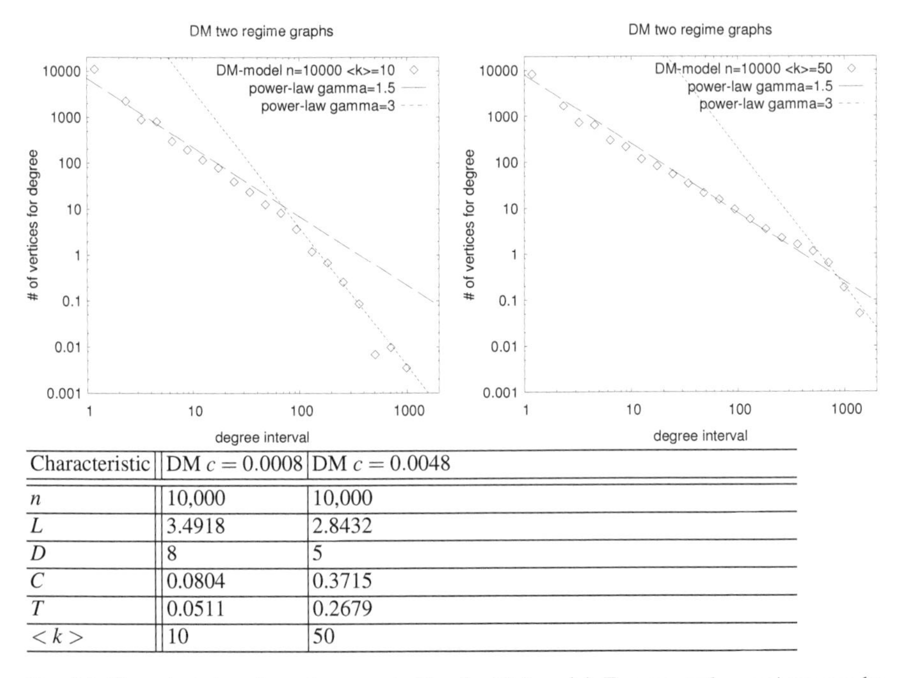

Characteristic	DM $c = 0.0008$	DM $c = 0.0048$
n	10,000	10,000
L	3.4918	2.8432
D	8	5
C	0.0804	0.3715
T	0.0511	0.2679
$<k>$	10	50

Fig. 2.8 Characteristics of graphs generated by the DM-model. Two power-law regimes can be observed. The higher the average degree $<k>$, the higher is the crossover degree k_{cross}

Model	ER	WS	BA	ST	(α,β)	DM
undirected	yes	yes	yes	yes	no	yes
directed	no	no	no	yes	yes	no
L	$\approx$5	$\approx$6	$\approx$4	$\approx$4	$\approx$4	$\approx$4
D	$\approx$7	$\approx$7	$\approx$6	$\approx$8	$\approx$9	$\approx$8
C	very low	high	low	high	high	high
T	very low	high	low	low	low	high
$C \lesseqgtr T$	$C = T$	$C = T$	$C = T$	$C \gg T$	$C \gg T$	$C > T$
$<k>$	10	10	10	10	$\approx$8	10
$P(k)$ tail	exp.	exp.	pl $\gamma = 3$	pl $\gamma = 3$ / $\gamma \in [2.5,\infty)$	pl $\gamma \in [2,\infty)$	pl $\gamma_1 = 1.5$ $\gamma_2 = 3$

Table 2.3 Comparison of graph models: ER-model, WS-model, BA-model, ST-model, (α,β)-model and DM-model for graphs with n=10,000 and $<k>= 10$. Approximate values are marked by $\approx$, pl denotes power-law

To compare the graph models discussed so far, Table 2.3 contrasts their characteristics. The ST-model, the (α,β)-model and the DM-model generate scale-free SWGs. Graphs generated by the ER-model and the BA-model exhibit a small clustering coefficient, ER-model and WS-model graphs are not scale-free.

Up to this point, only simple, i.e. unweighted models of SWGs have been discussed. As graph representations of natural phenomena are often weighted, extensions to weighted graphs are shortly mentioned in the following.

Vertex weights correspond to the importance of the represented entities, *edge weights* model the strength of interaction. If vertex weights are set to the degree of the vertex and edge weights to the product of the weight of its two vertices, a scale-free graph exhibits also a power-law distribution of vertex and edge weights, as observed in real-world data by Li and Chen [151] and Barrat et al. [17].

Barrat et al. [18] describe a generative model quite similar to the BA-model, only that vertex weights are the influencing factor for attachment rather than degree. When attaching a new vertex, the old vertex' weight is increased by the weight of the new vertex plus a parameter δ; for $\delta = 0$, this is equivalent to the BA-model. As $\delta \rightarrow \infty$, the power-law exponent decreases to $\gamma = 2$.

This chapter provided definitions and terminology of graph theory. Properties of Small World graphs were discussed and the most prominent graph generation models were reviewed. In the next chapter, SWGs occurring in natural language will be examined and a generation model for language will be developed.

2.2.9 Further Reading

A survey of models for the directed web graph that compares several models on various properties can be found in [40], including several models that not only model the addition of new vertices but also deletion of old ones. The works of Jon Kleinberg [141; 142, inter al.] convey the consequences for navigation and search algorithms when operating on web-like SWGs. Comprehensive surveys on SWGs in a World Wide Web context are [80], [55], for link prediction see [154]. On employing Small World network structures in distributed information retrieval, consult [127].

Arguments against over-estimating the findings of omnipresent large-scale properties are given by Keller [138]. While this author is right in the respect that the scale-free property found in many complex systems not necessarily implies a common architecture, it is still important to account for these properties when processing data exhibiting them.

Chapter 3
Small Worlds of Natural Language

Abstract In this chapter, power-law distributions and Small World Graphs originating from natural language data are examined in the fashion of Quantitative Linguistics. After giving several data sources that exhibit power-law distributions in rank-frequency in Section 3.1, graphs with Small World properties in language data are discussed in Section 3.2. We shall see that these characteristics are omnipresent in language data, and we should be aware of them when designing Structure Discovery processes. When knowing e.g. that a few hundreds of words make the bulk of words in a text, it is safe to use only these as contextual features without losing a lot of text coverage. Knowing that word co-occurrence networks possess the scale-free Small World property has implications for clustering these networks. An interesting aspect is whether these characteristics are only inherent to real natural language data or whether they can be produced with generators of linear sequences in a much simpler way than our intuition about language complexity would suggest — in other words, we shall see how distinctive these characteristics are with respect to tests deciding whether a given sequence is natural language or not. Finally, an emergent random text generation model that captures many of the characteristics of natural language is defined and quantitatively verified in Section 3.3.

3.1 Power-Laws in Rank-Frequency Distribution

G.K. Zipf [255, 256] described the following phenomenon: if all words in a corpus of natural language are arranged in decreasing order of frequency, then the relation between a word's frequency and its rank in the list follows a power-law. Since then, a significant amount of research has been devoted to the question how this property emerges and what kinds of processes generate such Zipfian distributions. Now, some datasets related to language will be presented that exhibit a power-law on their rank-frequency distribution. For this, the basic units as given in Section 1.3.1 will be examined.

3.1.1 Word Frequency

The relation between the frequency of a word at rank r and its rank is given by $f(r) \sim r^{-z}$, where z is the exponent of the power-law that corresponds to the slope of the curve in a log-log plot. The exponent z was assumed to be exactly 1 by Zipf; In natural language data, also slightly differing exponents in the range of about 0.7 to 1.2 are observed [see 252]. B. Mandelbrot [157] provided a formula that closer approximates the frequency distributions in language data, noticing that Zipf's law holds only for the medium range of ranks, whereas the curve is flatter for very frequent words and steeper for high ranks. Figure 3.1 displays the word rank-frequency distributions of corpora of different languages taken from the Leipzig Corpora Collection[1].

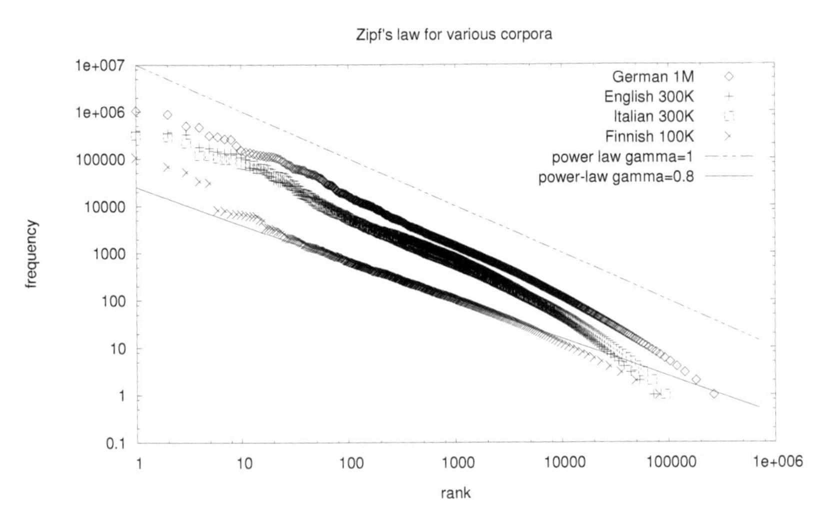

Fig. 3.1 Zipf's law for various corpora. The numbers next to the language give the corpus size in sentences. Enlarging the corpus does not effect the slope of the curve, but merely moves it upwards in the plot. Most lines are almost parallel to the ideal power-law curve with $z = 1$. Finnish exhibits a lower slope of $\gamma \approx 0.8$, akin to higher morphological productivity

There exist several exhaustive collections of research capitalising Zipf's law and related distributions[2] ranging over a wide area of datasets; here, only findings related to natural language are reported. A related distribution which will play a role at a later stage is the *lexical spectrum* [see 96], which gives the probability of choosing a word from the vocabulary with a given frequency. For natural language, the lexical spectrum follows a power-law with slope $\gamma = \frac{1}{z} + 1$, where z is the exponent of

[1] LCC, see http://www.corpora.uni-leipzig.de [July 7th, 2007].

[2] e.g. http://www.nslij-genetics.org/wli/zipf/index.html [April 1st, 2007] or http://linkage.rockefeller.edu/wli/zipf/index_ru.html [April 1st, 2007].

the Zipfian rank-frequency distribution. For the relation between lexical spectrum, Zipf's law and Pareto's law, see [2].

But Zipf's law in its original form is just the tip of the iceberg of power-law distributions in quantitative descriptions of language. While a Zipfian distribution for word frequencies can be obtained by a simple model of generating letter sequences with space characters as word boundaries [157; 177], these models based on "intermittent silence" can neither reproduce the distributions on sentence length [see 223], nor explain the relations of words in sequence. But before elaborating further on this point in Section 3.3, more power-law distributions in natural language are discussed and exemplified.

3.1.2 Letter N-grams

To continue with a counterexample, letter frequencies do not obey a power-law in the rank-frequency distribution. This also holds for letter N-grams (including the space character), yet for higher N, the rank-frequency plots show a large power-law regime with exponential tails for high ranks. Figure 3.2 shows the rank-frequency plots for letter N-grams up to $N = 6$ for the first 10,000 sentences of the British National Corpus (BNC[3], [50].).

Still, letter frequency distributions can be used to show that letters are not forming letter bigrams from single letters independently, but there are restrictions on their combination. While this intuitively seems obvious for letter combination, the following test is proposed for quantitatively examining the effects of these restrictions: from letter unigram probabilities, a text is generated that follows the letter unigram distribution by randomly and independently drawing letters according to their distribution and concatenating them. The letter bigram frequency distribution of this generated text can be compared to the letter bigram frequency distribution of the real text the unigram distribution was measured from. Figure 3.3 shows the generated and the real rank-frequency plot, again from the small BNC sample.

The generated bigrams without restrictions predict a higher number of different bigrams and lower frequencies for bigrams of high ranks as compared to the real text bigram statistics. This shows that letter combination restrictions do exist, as not nearly all bigrams predicted by the generation process were observed, resulting in higher counts for valid bigrams in the sample.

3.1.3 Word N-grams

For word N-grams, the relation between rank and frequency follows a power-law, just as for words (unigrams). Figure 3.4 (left) shows the rank-frequency plots up to

[3] http://www.natcorp.ox.ac.uk/ [April 1st, 2007]

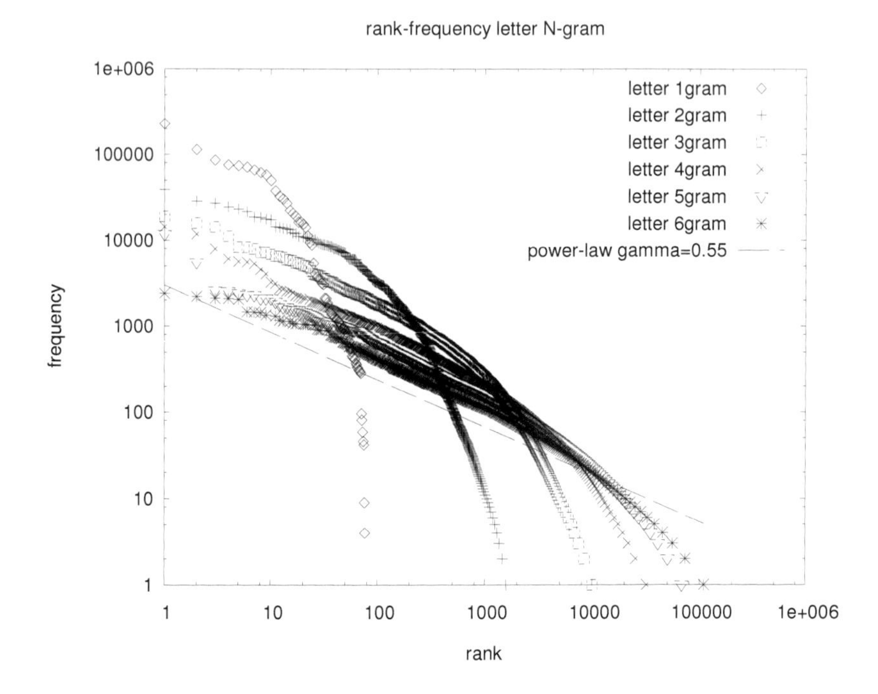

Fig. 3.2 Rank-Frequency distributions for letter N-grams for the first 10,000 sentences in the BNC. Letter N-gram rank-frequency distributions do not exhibit power-laws on the full scale, but increasing N results in a larger power-law regime for low ranks

$N = 4$, based on the first 1 million sentences of the BNC. As more different word combinations are possible with increasing N, the curves get flatter as the same total frequency is shared amongst more units, as previously observed by e.g. [226] and [117]. Testing concatenation restrictions quantitatively as above for letters, it might at the first glance seem surprising that the curve for a text generated with word unigram frequencies differs only very little from the word bigram curve, as Figure 3.4 (right) shows. Small differences are only observable for low ranks: more top-rank generated bigrams reflect that words are usually not repeated in the text. More low-ranked and less high-ranked real bigrams indicate that word concatenation takes place not entirely without restrictions, yet is subject to much more variety than letter concatenation. This coincides with the intuition that it is, for a given word pair, almost always possible to form a correct English sentence in which these words are neighbours. Regarding quantitative aspects, the frequency distribution of word bigrams can be produced by a generation process based on word unigram probabilities. In Section 3.3, a measure will be introduced that can better distinguish between a text generated in this way and a real text.

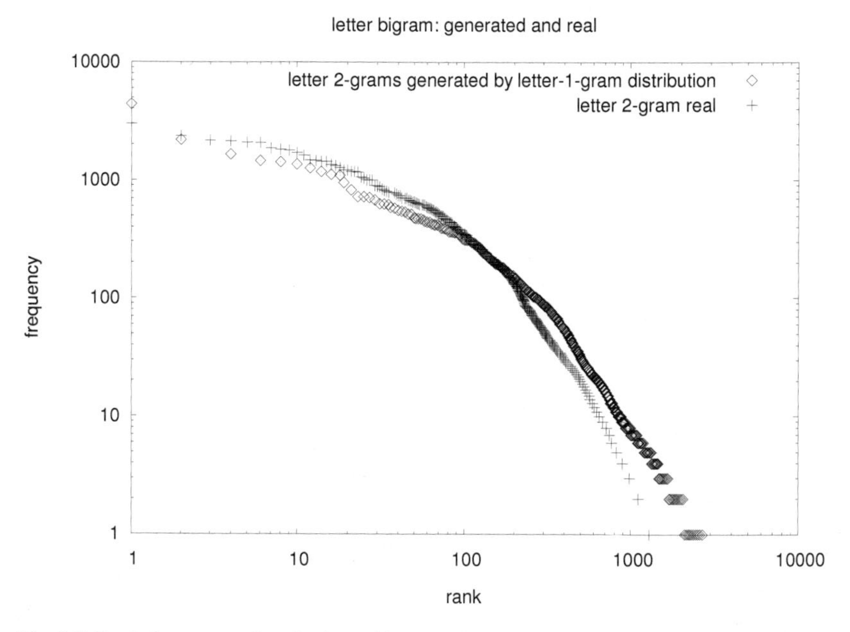

Fig. 3.3 Rank-frequency plots for letter bigrams, for a text generated from letter unigram probabilities and for the BNC sample

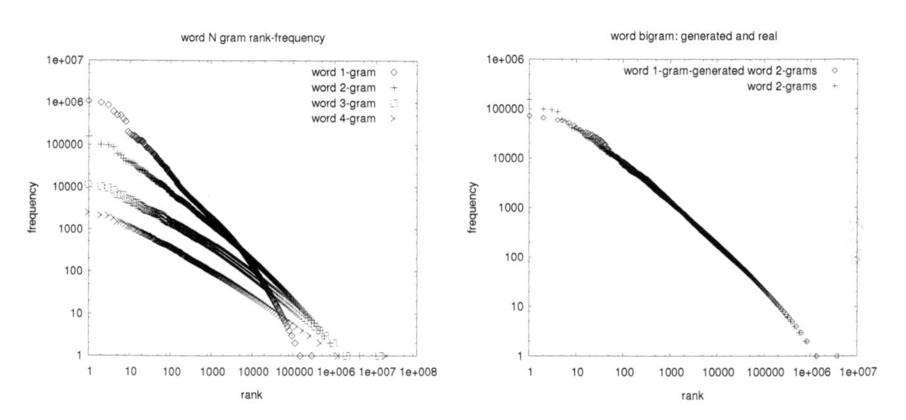

Fig. 3.4 Left: rank-frequency distributions for word N-grams for the first 1 million sentences in the BNC. Word N-gram rank-frequency distributions exhibit power-laws. (Right: rank-frequency plots for word bigrams, for a text generated from letter unigram probabilities and for the BNC sample

3.1.4 Sentence Frequency

In larger corpora that are compiled from a variety of sources, a considerable number of duplicate sentences can be found. In the full BNC, which serves as data basis for examining this, 7.3% of the sentences occur two or more times. The most frequent sentences are "Yeah.", "Mm.", "Yes." and "No.", which are mostly found in the section of spoken language. But also longer expressions like "Our next bulletin is at 10.30 p.m." have a count of over 250. The sentence frequencies also follow a power-law with an exponent close to 1 (cf. Figure 3.5), indicating that Zipf's law also holds for sentence frequencies.

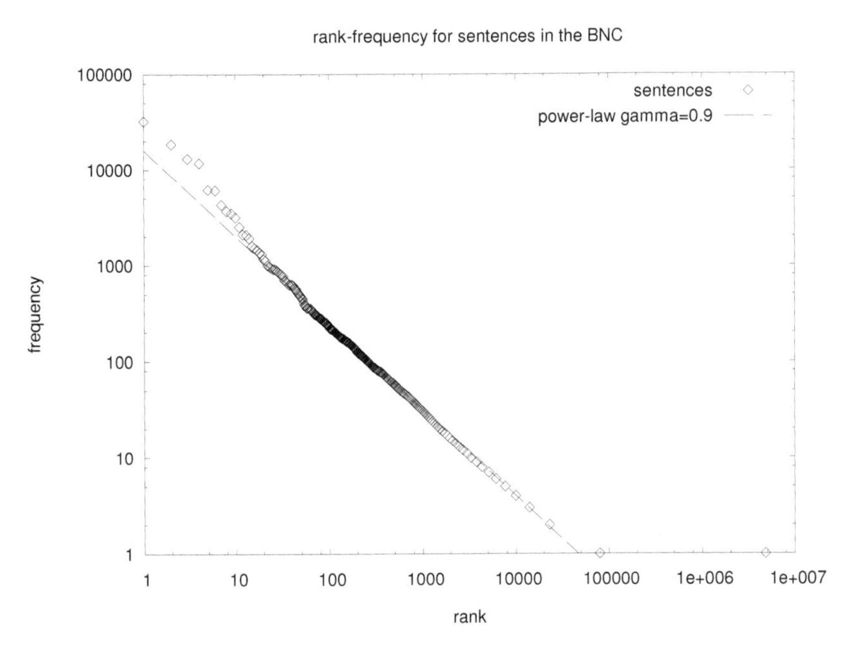

Fig. 3.5 Rank-frequency plot for sentence frequencies in the full BNC, following a power-law with $\gamma \approx 0.9$, but with a high fraction of sentences occurring only once

3.1.5 Other Power-Laws in Language Data

The results above strongly suggest that when counting document frequencies in large collections such as the World Wide Web, another power-law distribution would be found, but an analysis has not been carried out and would require access to the index of a web search engine.

Further, there are more power-laws in language-related areas, some of which are provided briefly to illustrate their omnipresence:

- Web page requests follow a power-law, which was employed for a caching mechanism in [110].
- Related to this, frequencies of web search queries during a fixed time span also follow a power-law, as exemplified in Figure 3.6 for a 7 million queries log of AltaVista[4] as used by Lempel and Moran [149].
- The number of authors of Wikipedia[5] articles was found to follow a power-law with $\gamma \approx 2.7$ for a large regime in [242], who also discusses other power-laws regarding the number of links.

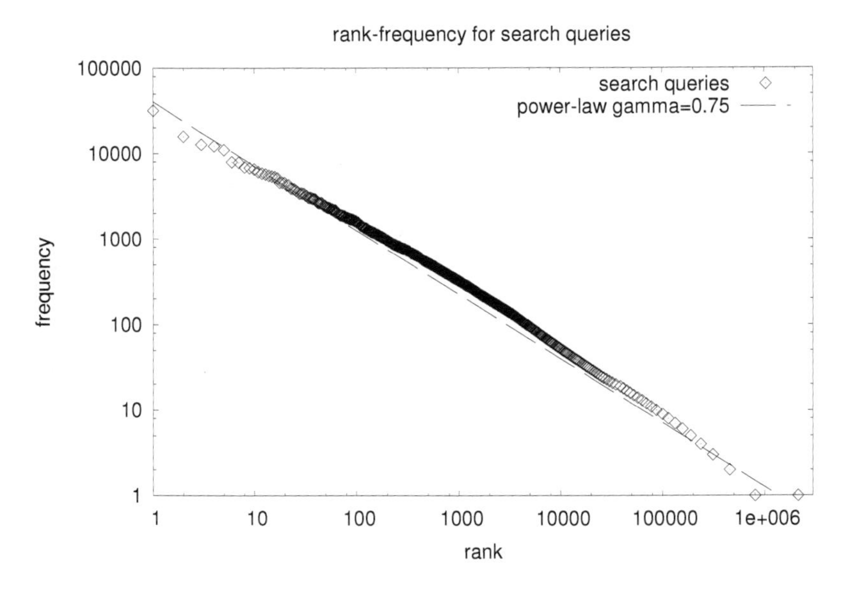

Fig. 3.6 Rank-frequency plot for AltaVista search queries, following a power-law with $\gamma \approx 0.75$

3.1.6 Modelling Language with Power-Law Awareness

While simple language models such as the N-gram model produce power-laws in the rank-frequency distribution, this does not necessarily hold for generative models such as Latent Dirichlet Allocation (LDA, [35]) or other graphical models [133]. Such models, on the other hand have shown to be very useful for modelling natural

[4] http://www.altavista.com [August 31, 2011]

[5] http://www.wikipedia.org [August 31, 2011]

language semantics, and possess better smoothing capabilities and a stronger mathematical foundation than simple N-gram models. It is therefore desirable to utilise the strengths of these generative models, without suffering from their deficiencies regarding the power-law distributions. This is carried out in the two-stage language model of Goldwater et al. [113], which is highly relevant for the issues presented in this chapter. Here, arbitrary generative processes that do not necessarily adhere to power-law distributions are adapted to match this distribution using a stochastic adaptation process that 'corrects' the frequency distribution of the generative model in order to produce a power-law. This sheds light on the interaction between types and tokens, and provides justification for the success of term weighting schemes that dampen the raw word frequencies by e.g. taking the logarithm, such as *tf-idf* [cf. 13].

3.2 Scale-Free Small Worlds in Language Data

Whereas the previous section discussed the shape of rank-frequency distributions for natural language units, now the properties of graphs with units represented as vertices and relations between them as edges will be in the focus of interest. As already stated in Section 1.3.2, internal as well as contextual features can be employed for computing similarities between language units that are represented as (possibly weighted) edges in the graph. Some of the graphs discussed here can be classified in being scale-free Small World graphs; others differ from these characteristics and represent other, but related graph classes.

3.2.1 Word Co-occurrence Graph

The notion of *word co-occurrence* is used to model dependencies between words. If two words X and Y occur together in some contextual unit of information (as neighbours, in a word window of 5, in a clause, in a sentence, in a paragraph), they are said to co-occur. When regarding words as vertices and edge weights as the number of times two words co-occur, the *word co-occurrence graph* of a corpus is given by the entirety of all word co-occurrences.

In the following, specifically two types of co-occurrence graphs are considered: the graph induced by the word neighbour relation, henceforth called neighbour-based graph, and the graph induced by sentence-based word co-occurrence, henceforth called sentence-based graph. The neighbour-based graph can be undirected or directed with edges going from left to right words as found in the corpus, the sentence-based graph is undirected. To illustrate co-occurrence graphs, Figure 3.7 displays the sentence-based co-occurrence graph and the co-occurrence graph based on neighbouring words for a song by Radiohead.

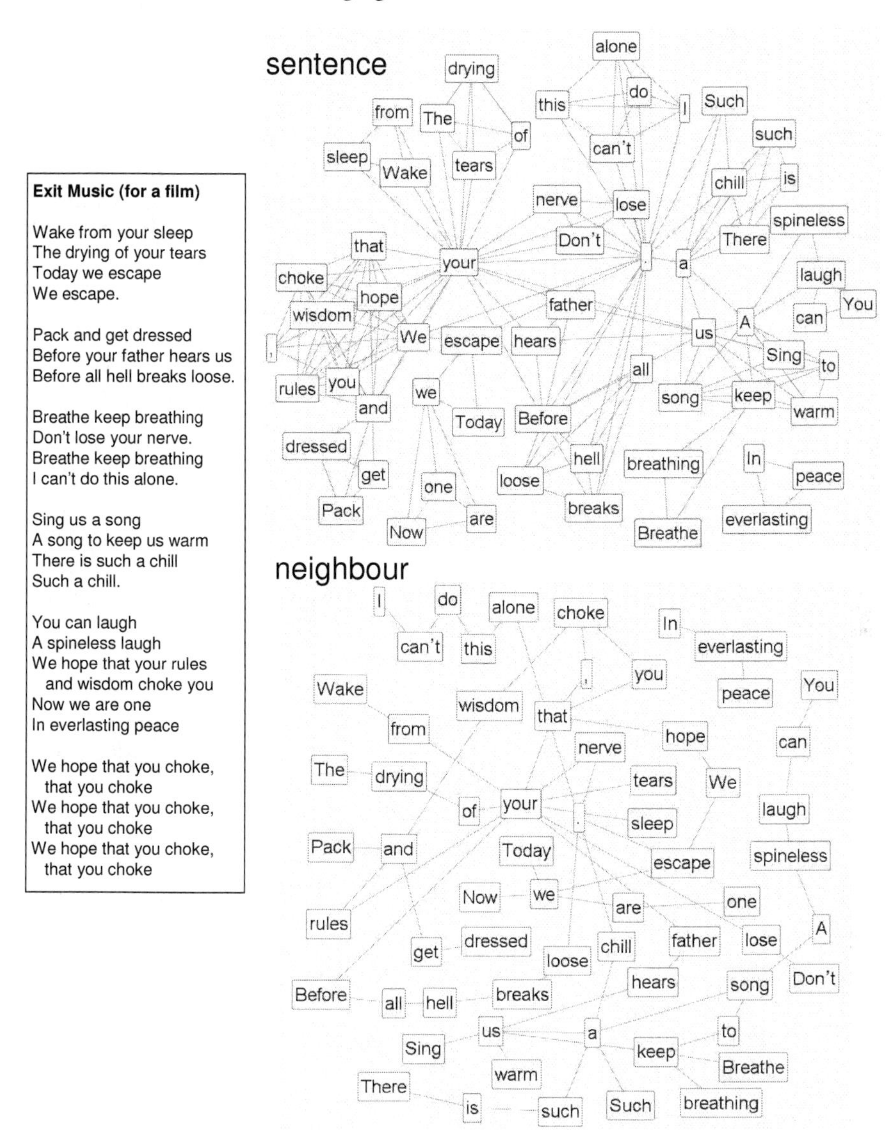

Fig. 3.7 The co-occurrence graphs for the song "Exit Music (for a film)" by Radiohead. Upper graph: words co-occurring in the same verse. Lower graph: words co-occurring as neighbours. Edge weights are omitted. Notice that the neighbouring relation does not cross verse boundaries

To find out whether the co-occurrence of two specific words A and B is merely due to chance or exhibits a statistical dependency, measures are used that compute, to what extent the co-occurrence of A and B is statistically significant. Many significance measures can be found in the literature, for extensive overviews consult e.g. [94] or [42]. In general, the measures compare the probability of A and B to co-occur under the assumption of their statistical independence with the number of times A and B actually co-occurred in the corpus. In this work, the log likelihood ratio [86] is used to sort the chaff from the wheat. It is given in expanded form by Bordag [42]:

$$-2\log\lambda = 2\begin{bmatrix} n\log n - n_A\log n_A - n_B\log n_B + n_{AB}\log n_{AB} \\ +(n-n_A-n_B+n_{AB})\log(n-n_A-n_B+n_{AB}) \\ +(n_A-n_{AB})\log(n_A-n_{AB})+(n_B-n_{AB})\log(n_B-n_{AB}) \\ -(n-n_A)\log(n-n_A)-(n-n_B)\log(n-n_B) \end{bmatrix} \quad (3.1)$$

where n is the total number of contexts, n_A the frequency of A, n_B the frequency of B and n_{AB} the number of co-occurrences of A and B. As pointed out by Moore [180], this formula overestimates the co-occurrence significance for small n_{AB}. For this reason, often a frequency threshold t on n_{AB} (e.g. a minimum of $n_{AB} = 2$) is applied. Further, a significance threshold s regulates the density of the graph; for the log likelihood ratio, the significance values correspond to the χ^2 tail probabilities [180], which makes it possible to translate the significance value into an error rate for rejecting the independence assumption. For example, a log likelihood ratio of 3.84 corresponds to a 5% error in stating that two words do not occur by chance, a significance of 6.63 corresponds to 1% error.

The operation of applying a significance test results in pruning edges being in existence due to random noise and keeping almost exclusively edges that reflect a true association between their endpoints. Graphs that contain all significant co-occurrences of a corpus, with edge weights set to the significance value between its endpoints, are called *significant co-occurrence graphs* in the remainder. For convenience, no singletons in the graph are allowed, i.e. if a vertex is not contained in any edge because none of the co-occurrences for the corresponding word is significant, then the vertex is excluded from the graph.

As observed by Ferrer-i-Cancho and Solé [95] and Quasthoff et al. [198], word co-occurrence graphs exhibit the scale-free Small World property. This goes in line with co-occurrence graphs reflecting human associations [see 201] and human associations in turn forming Small World graphs [see 232]. The claim is confirmed here on an exemplary basis with the graph for LCC's 1 million sentence corpus for German. Figure 3.8 gives the degree distributions and graph characteristics for various co-occurrence graphs.

The shape of the distribution is dependent on the language, as Figure 3.9 shows. Some languages — here English and Italian — have a hump-shaped distribution in the log-log plot where the first regime follows a power-law with a lower exponent than the second regime, as observed by Ferrer-i-Cancho and Solé [95]. For the Finnish and German corpora examined here, this effect could not be found in the

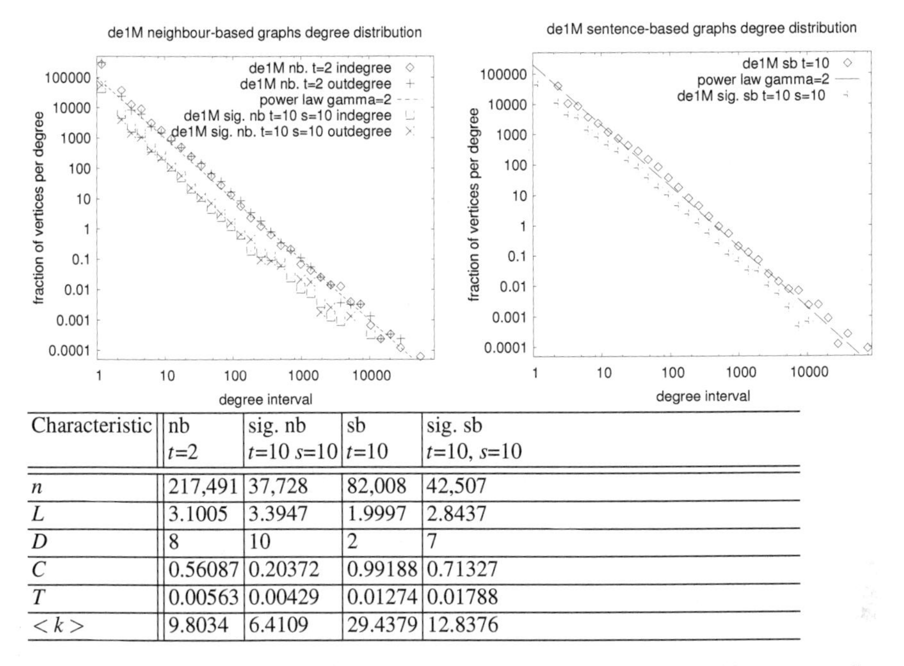

Characteristic	nb t=2	sig. nb t=10 s=10	sb t=10	sig. sb t=10, s=10
n	217,491	37,728	82,008	42,507
L	3.1005	3.3947	1.9997	2.8437
D	8	10	2	7
C	0.56087	0.20372	0.99188	0.71327
T	0.00563	0.00429	0.01274	0.01788
$<k>$	9.8034	6.4109	29.4379	12.8376

Fig. 3.8 Graph characteristics for various co-occurrence graphs of LCC's 1 million sentence German corpus. Abbreviations: nb = neighbour-based, sb = sentence-based, sig. = significant, t = co-occurrence frequency threshold, s = co-occurrence significance threshold. While the exact shapes of the distributions are language and corpus dependent, the overall characteristics are valid for all samples of natural language of sufficient size. The slope of the distribution is invariant to changes of thresholds. Characteristic path length and a high clustering coefficient at low average degrees are characteristic for SWGs

data. This property of two power-law regimes in the degree distribution of word co-occurrence graphs motivated the DM-model (see Section 2.2.7, [83]). There, the cross-over-point of the two power-law regimes is motivated by a so-called *kernel lexicon* of about 5,000 words that can be combined with all words of a language.

The original experiments of Ferrer-i-Cancho and Solé [95] operated on a word co-occurrence graph with window size 2: an edge is drawn between words if they appear together at least once in a distance of one or two words in the corpus. Reproducing their experiment with the first 70 million words of the BNC and corpora of German, Icelandic and Italian of similar size reveals that the degree distribution of the English and the Italian graph is in fact approximated by two power-law regimes. In contrast to this, German and Icelandic show a single power-law distribution, just as in the experiments above, see Figure 3.10. These results suggest that two power-law regimes in word co-occurrence graphs with window size 2 are not a language universal, but only hold for some languages.

To examine the hump-shaped distributions further, Figure 3.11 displays the degree distribution for the neighbour-based word co-occurrence graphs and the word-co-occurrence graphs for connecting only words that appear in a distance of 2. It

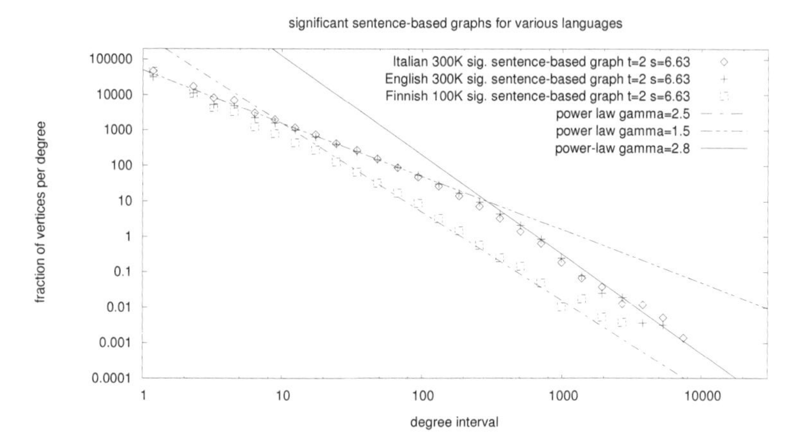

Fig. 3.9 Degree distribution of significant sentence-based co-occurrence graphs of similar thresholds for Italian, English and Finnish

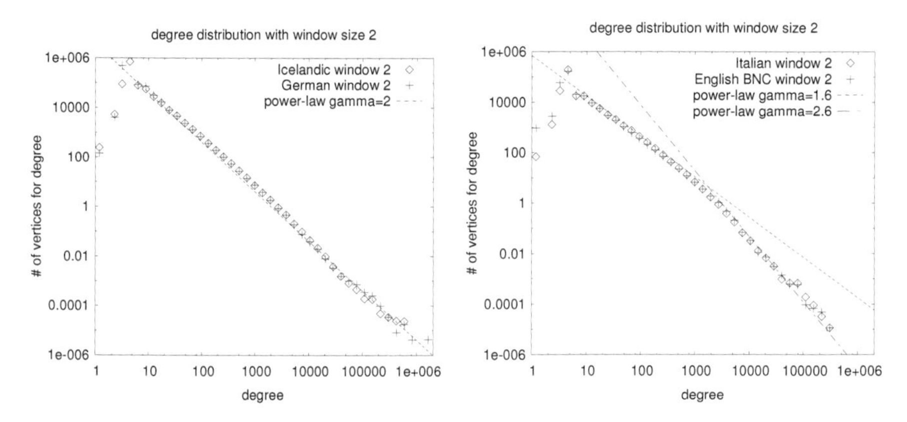

Fig. 3.10 Degree distributions in word co-occurrence graphs for window size 2. Left: the distribution for German and Icelandic is approximated by a power-law with $\gamma = 2$. Right: for English (BNC) and Italian, the distribution is approximated by two power-law regimes

becomes clear from the plots that the hump-shaped distribution is mainly caused by words co-occurring in distance 2, whereas the neighbour-based graph shows only a slight deviation from a single power-law. Together with the observations from sentence-based co-occurrence graphs of different languages in Figure 3.9, it is obvious that a hump-shaped distribution with two power-law regimes is caused by long-distance relationships between words, if present at all.

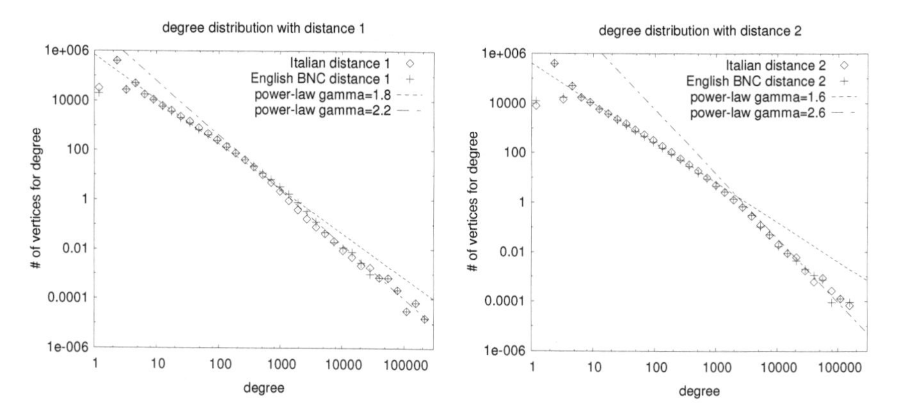

Fig. 3.11 Degree distributions in word co-occurrence graphs for distance 1 and distance 2 for English (BNC) and Italian. The hump-shaped distribution is much more distinctive for distance 2

3.2.1.1 Applications of Word Co-occurrences

Word co-occurrence statistics are an established standard and have been used in many language processing systems. The author has used word co-occurences in many tasks, including bilingual dictionary acquisition as in [75], semantic lexicon extension [see 33] and visualisation of concept trails in [24]. In Chapters 5, 6 and 7 they will be used as building blocks in various Structure Discovery processes.

3.2.2 Co-occurrence Graphs of Higher Order

The significant word co-occurrence graph of a corpus represents words that are likely to appear near to each other. When interested in words co-occurring with similar other words, it is possible to transform the above defined (first order) co-occurrence graph into a second order co-occurrence graph by drawing an edge between two words A and B if they share a common neighbour in the first order graph. Whereas the first order word co-occurrence graph represents the global context per word, the corresponding second order graph contains relations between words which have similar global contexts. The edge can be weighted according to the number of common neighbours, e.g. by $weight = |neigh(A) \cap neigh(B)|$. Figure 3.12 shows neighbourhoods of the significant sentence-based first-order word co-occurrence graph from LCC's English web corpus[6] for the words *jazz* and *rock*. Taking into account only the data depicted, *jazz* and *rock* are connected with an edge of weight 3 in the second-order graph, corresponding to their common neighbours *album*, *music* and *band*. The fact that they share an edge in the first order graph is ignored.

[6] http://corpora.informatik.uni-leipzig.de/?dict=en [April 1st, 2007]

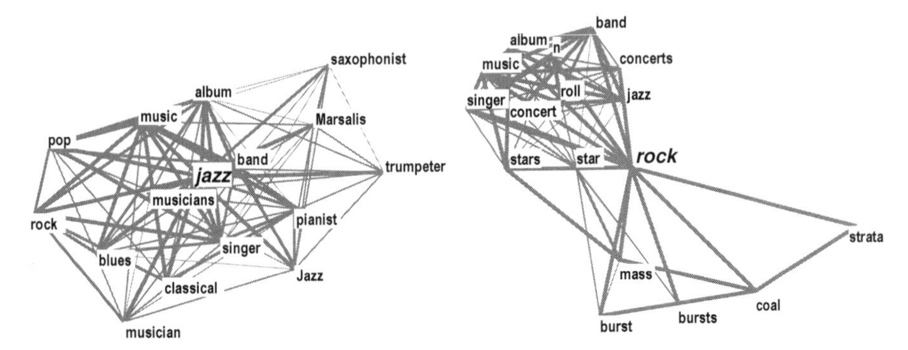

Fig. 3.12 Neighbourhoods of *jazz* and *rock* in the significant sentence-based word co-occurrence graph as displayed on LCC's English Corpus website. Both neighbourhoods contain *album*, *music* and *band*, which leads to an edge weight of 3 in the second-order graph

In general, a graph of order $N+1$ can be obtained from the graph of order N, using the same transformation. The higher order transformation without thresholding is equivalent to a multiplication of the unweighted adjacency matrix A with itself, then zeroing the main diagonal by subtracting the degree matrix of A. Since the average path length of scale-free SW graphs is short and local clustering is high, this operation leads to an almost fully connected graph in the limit, which does not allow to draw conclusions about the initial structure. Thus, the graph is pruned in every iteration N in the following way: For each vertex, only the max_N outgoing edges with the highest weights are taken into account. Notice that this vertex degree threshold max_N does not limit the maximum degree, as thresholding is asymmetric. This operation is equivalent with only keeping the max_N largest entries per row in the adjacency matrix $A = (a_{ij})$, then $A_t = (sign(a_{ij} + a_{ji}))$, resulting in an undirected graph.

To examine quantitative effects of the higher order transformation, the sentence-based word co-occurrence graph of LCC's 1 million German sentence corpus ($s = 6.63$, $t = 2$) underwent this operation. Figure 3.13 depicts the degree distributions for $N = 2$ and $N = 3$ for different max_N.

Applying the max_N threshold causes the degree distribution to change, especially for high degrees. In the third order graph, two power-law regimes are observable.

Studying the degree distribution of higher order word co-occurrence graphs reveals that the characteristic of being governed by power-laws is invariant to the higher order transformation, yet the power-law exponent changes. This indicates that the power-law characteristic is inherent at many levels in natural language data.

To examine what this transformation yields on the graphs generated by the random graph models in the previous chapter, Figure 3.14 shows the degree distribution of second order and third order graphs as generated by the models described in Section 2.2. The underlying first order graphs are the undirected graphs of order 10,000 and size 50,000 ($<k>= 10$) from the BA-model, the ST-model, and the DM-model.

While the thorough interpretation of second order graphs of random graphs might be subject to further studies, the following should be noted: the higher order trans-

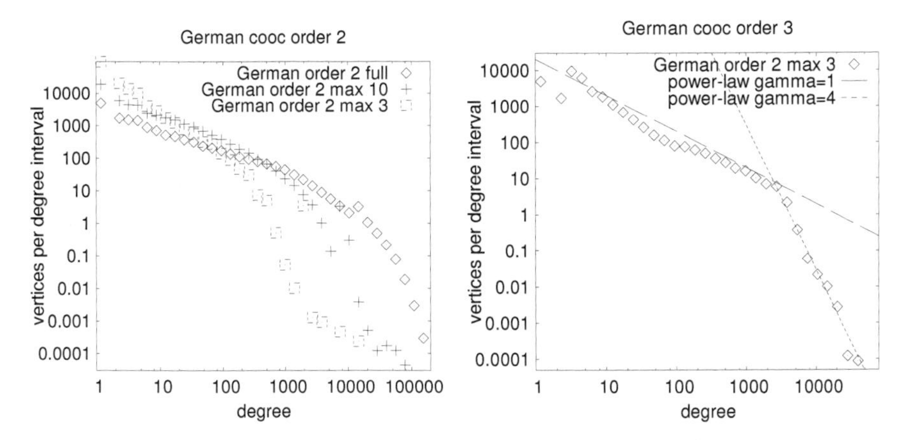

Fig. 3.13 Degree distributions of word-co-occurrence graphs of higher order. The first order graph is the sentence-based word co-occurrence graph of LCC's 1 million German sentence corpus ($s = 6.63$, $t = 2$). Left: $N = 2$ for $max_2 = 3$, $max_2 = 10$ and $max_2 = \infty$. Right: $N = 3$ for $t_2 = 3$, $t_3 = \infty$

formation reduces the power-law exponent of the BA-model graph from $\gamma = 3$ to $\gamma = 2$ in the second order and to $\gamma \approx 0.7$ in the third order. For the ST-model, the degree distribution of the full second order graph shows a maximum around $2M$, then decays with a power-law with exponent $\gamma \approx 2.7$. In the third order ST-graph, the maximum moves to around $4M^2$ for sufficient max_2. The DM-model second order graph shows, like the first order DM-model graph, two power-law regimes in the full version, and a power-law with $\gamma \approx 2$ for the pruned versions. The third order degree distribution exhibits many more vertices with high degrees than predicted by a power-law.

In summary, all random graph models exhibit clear differences to word co-occurrence networks with respect to the higher order transformation. The ST-model shows maxima depending on the average degree of the first oder graph. The BA-model's power law is decreased with higher orders, but is able to explain a degree distribution with power-law exponent 2. The full DM model exhibits the same two power-law regimes in the second order as observed for German sentence-based word co-occurrences in the third order.

3.2.2.1 Applications of Co-occurrence Graphs of Higher Orders

In [25] and [156], the utility of word co-occurrence graphs of higher orders are examined for lexical semantic acquisition. The highest potential for extracting paradigmatic semantic relations can be attributed to second- and third-order word co-occurrences. Second-order graphs are evaluated by Bordag [42] against lexical semantic resources. In Section 6, a second-order graph based on neighbouring words on both sides will be, amongst other mechanisms, used for inducing syntactic word classes.

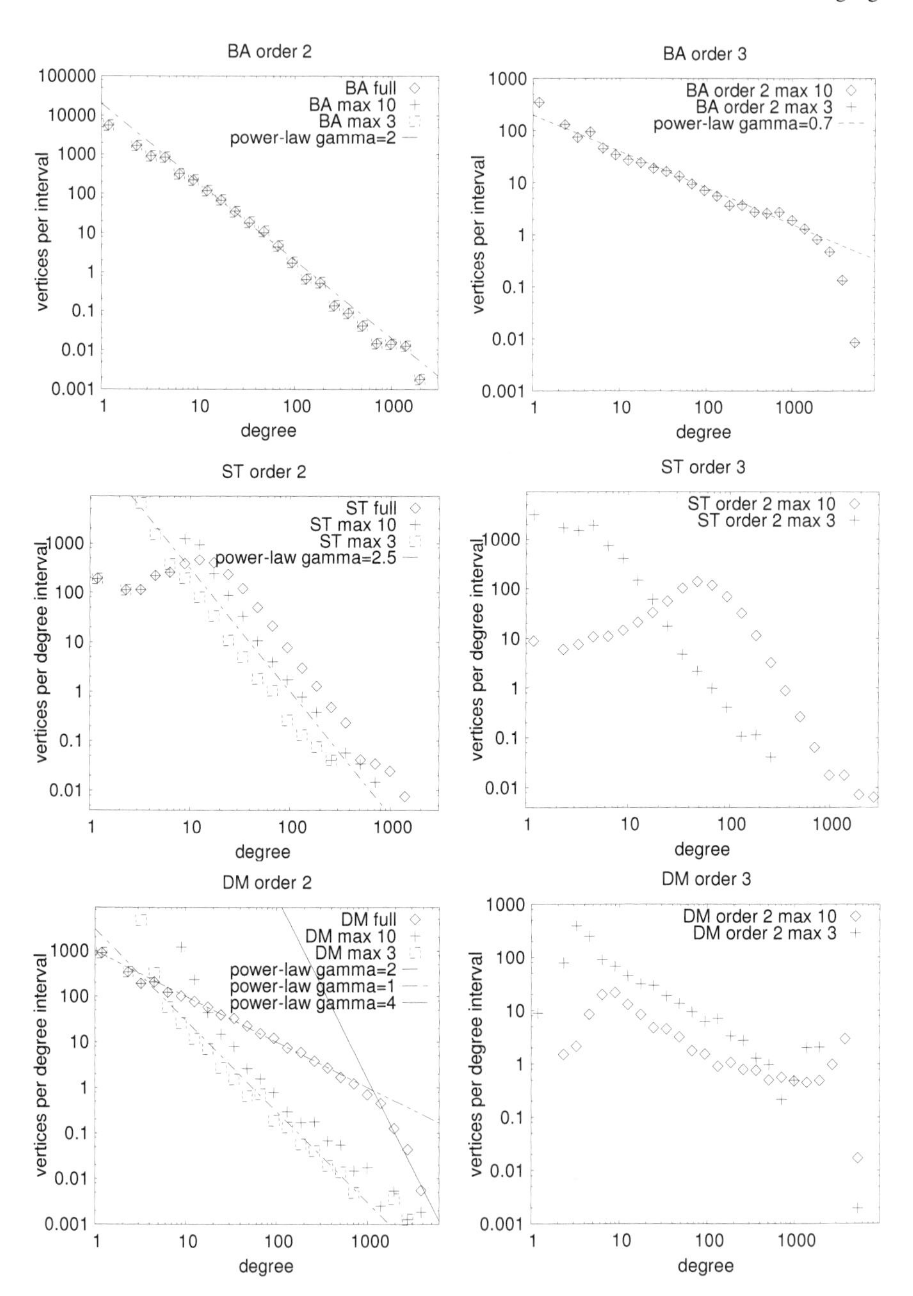

Fig. 3.14 Second and third order graph degree distributions for BA-model, ST-model and DM-model graphs

3.2.3 Sentence Similarity

Using words as internal features, the similarity of two sentences can be measured by how many common words they share. Since the few top frequency words are contained in most sentences as a consequence of Zipf's law, their influence should be downweighted or they should be excluded to arrive at a useful measure for sentence similarity. Here, the sentence similarity graph of sentences sharing at least two common words is examined, with the maximum frequency of these words bounded by 100. The corpus of examination is here LCC's 3 million sentences of German. Figure 3.15 shows the component size distribution for this sentence-similarity graph, Figure 3.16 shows the degree distributions for the entire graph and for its largest component.

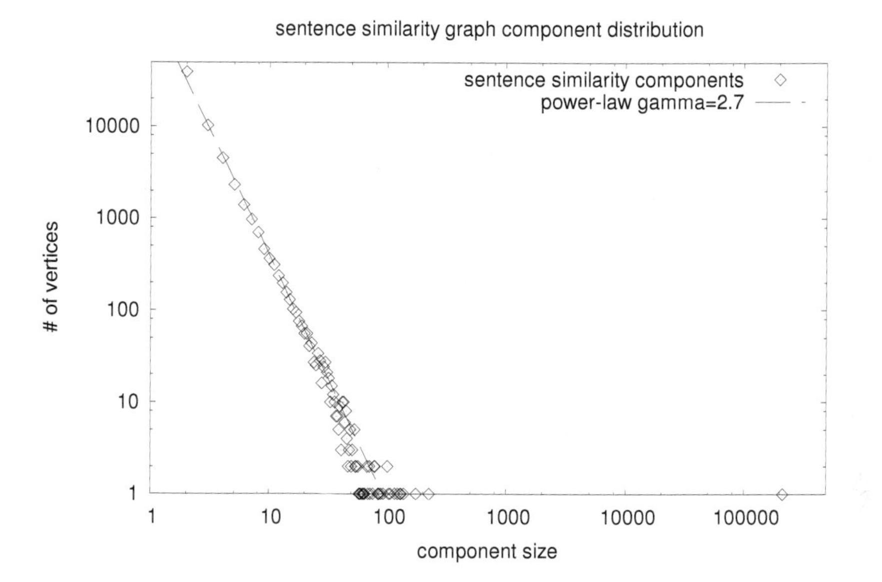

Fig. 3.15 Component size distribution for the sentence similarity graph of LCC's 3 million sentence German corpus. The component size distribution follows a power-law with $\gamma \approx 2.7$ for small components, the largest component comprises 211,447 out of 416,922 total vertices. The component size distribution complies to the theoretical results of Aiello et al. [8], see Section 2.2.8

The degree distribution of the entire graph follows a power-law with γ close to 1 for low degrees and decays faster for high degrees, the largest component's degree distribution plot is flatter for low degrees. This resembles the broad-scale degree distribution (cf. Section 2.2.4) and can be attributed to limited sentence length: as sentences are not arbitrarily long (see more on sentence length distribution in Section 3.3.6), they cannot be similar to an arbitrary high number of other sentences with respect to the measure discussed here, as the number of sentences per feature word is bound by the word frequency limit. Cost in terms of Section 2.2.4 corre-

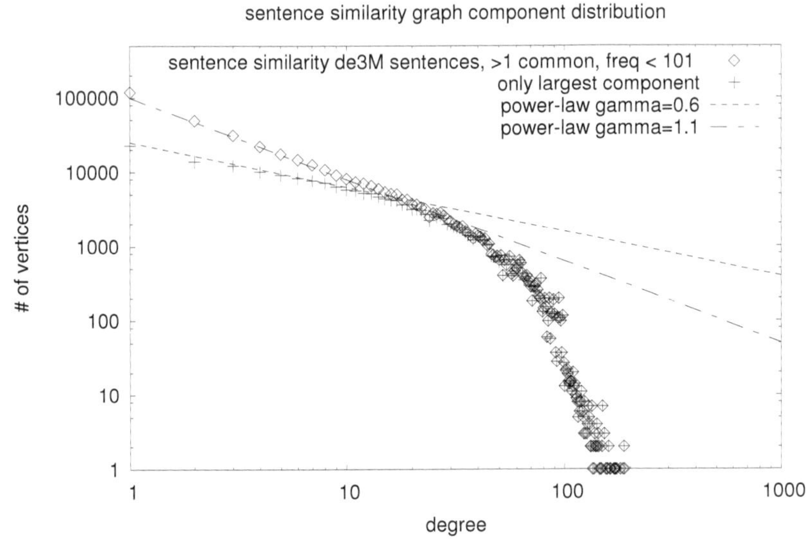

Characteristic	Sentence Similarity Graph	Largest Component Only
n	416,922	211,447
L	11.4850	11.4850
D	34	34
C	0.8836	0.8812
T	0.8807	0.8769
$<k>$	10.5968	17.3982

Fig. 3.16 Degree distribution for the sentence similarity graph of LCC's 3 million sentence German corpus and its largest component. An edge between two vertices representing sentences is drawn if the sentences share at least two words with corpus frequency <101, singletons are excluded

sponds to sentence length restrictions here. However, the extremely high values for transitivity and clustering coefficient and the low γ values for the degree distribution for low degree vertices and comparably long average shortest path lengths indicate that the sentence similarity graph belongs to a different graph class than the ones presented in Section 2.2.

3.2.3.1 Applications of the Sentence Similarity Graph

A similar measure is used in [26] for document similarity, obtaining well-correlated results when evaluating against a given document classification. A precision-recall tradeoff arises when lowering the frequency threshold for feature words or increasing the minimum number of common feature words two documents must have in order to be connected in the graph: both improves precision but results in many singleton vertices which lowers the total number of documents that are considered.

3.2.4 Summary on Scale-Free Small Worlds in Language Data

The examples above confirm the claim that graphs built on various aspects of natural language data often exhibit the scale-free Small World property or similar characteristics. Experiments with generated text corpora suggest that this is mainly due to the power-law frequency distribution of language units, cf. also Section 3.3. The slopes of the power-law approximating the degree distributions can often not be produced using the random graph generation models of Section 2.2: especially, all previously discussed generation models fail on explaining the properties of word co-occurrence graphs, where $\gamma \approx 2$ was observed as power-law exponent of the degree distribution. Of the generation models inspired by language data, the ST-model exhibits $\gamma = 3$, whereas the universality of the DM-model to capture word co-occurrence graph characteristics can be doubted after examining data from different languages.

Therefore, a new generation model is described in the following section, which succeeds in modelling the co-occurrence graph and other properties of natural language. For this, the graph is not generated directly, but obtained from a linear sequence of generated text.

3.3 An Emergent Random Generation Model for Language

Now, a graph-based emergent generation model for language will be given that reproduces the large-scale behaviour of language, i.e. a Zipfian distribution of word frequency and a power-law distribution with $\gamma = 2$ in the co-occurrence graph. Further, the model will produce sentence-like concatenations of words. At this, these characteristics emerge in a generation process, rather than being directly constrained by them. The model is inspired by SWG generation models as discussed in Section 2.2 and reproduces the language-characteristic graphs discussed above. This is the main contribution of the generation model and its distinguishing feature from previous random text models. The random text model first appeared in [28].

The study of random text models unveils, which of the characteristics of natural language can be produced by well-understood mechanisms. When keeping these mechanisms as simple as possible, they can be perceived as plausible explanations for the origin of language structure. Common characteristics of random text and natural language indicate the amount of structure that is explained by the generation process, differences indicate missing aspects in the random generator. These discrepancies can inspire both more complex random text models and Structure Discovery processes that employ the yet unmodelled regularities that cause them.

Before diving into the details, the following clarification must be stressed: emergent random text models differ from the commonly used generative models (such as Hidden Markov Models [e.g. 57] or Graphical Models [133]) in the important aspect that emergent models are not trained on existing data (be it annotated or raw), but define a process to generate random text that adheres to quantitative properties, such as the ones discussed in the previous sections of this chapter. Thus, they do not model real language data, but rather explain the origins of these quantitative properties. Knowledge about these properties can inform the construction of language models, e.g. in the form of priors in Bayesian learning (cf. [133; 113])).

3.3.1 Review of Emergent Random Text Models

B. Mandelbrot [157] provided a word generation model that produces random words of arbitrary average length in the following way: with a probability w, a word separator is generated at each step, with probability $(1-w)/N$, a letter from an alphabet of size N is generated, each letter having the same probability. This is sometimes called the "monkey at the typewriter" [177]. The frequency distribution follows a power-law for long streams of words, yet the equiprobability of letters causes the plot to show a step-wise rather than a smooth behaviour, as examined in [96] and shown in Figure 3.18. In the same study, a smooth rank distribution could be obtained by setting the letter probabilities according to letter frequencies in a natural language text. But the question of how these letter probabilities emerge remained unanswered[7]. Li [152] formally proves that the power-law frequency distribution in this random text model is a consequence of the transformation from the word's length to its rank, stretching the exponential distribution to a power-law.

Another random text model is given in [225], which does not take an alphabet of single letters into consideration. Instead, at each time step, a previously unseen new word is added to the stream with probability α, whereas with probability $(1-\alpha)$, the next word is chosen amongst the words at previous positions. As words with higher frequency in the already generated stream have a higher probability of being added again, this imposes a strong competition among different words, resulting in a frequency distribution that follows a power-law with exponent $\gamma = (1-\alpha)$. This was taken up by Zanette and Montemurro [252], who slightly modify Simon's model. They introduce sublinear vocabulary growth by additionally making the new word probability dependent onto the time step. Furthermore, they apply a threshold on the maximal probability a previously seen word is generated with, being able to modify the exponent z as well as to model the flatter curve for high frequency words.

Neither the Mandelbrot nor the Simon generation model takes the sequence of words into account. Simon treats the previously generated stream as a bag of words, and Mandelbrot does not consider the previous stream at all. This is certainly an over-simplification, as natural language exhibits structural properties within sentences and texts that are not grasped by a bag of words approach.

The framework of Goldwater et al. [113] (see Section 3.1.6) allows to produce power-laws for word frequencies with varying exponents. Due to its generality, it could incorporate state-of-the-art generative machine learning models that address word order, see [243].

---

[7] A suitable random initialisation of letter frequencies can produce a smooth rank distribution, yet is still not emergent.

The work by Kanter and Kessler [135] is, as far as the author is aware, the only piece of research that explicitly addresses word order awareness for random text generation. They show that a 2-parameter Markov process gives rise to a stationary distribution that exhibits the word or letter frequency distribution characteristics of natural language. However, the Markov process is initialised such that any state has exactly two successor states, which means that after each word only two different succeeding words are possible. This is certainly not reflecting natural language properties, where in fact successor frequencies of words follow a power-law and more successors can be observed for more frequent words. But even when allowing a more realistic number of successor states, the transition probabilities of a Markov model need to be initialised a priori in a sensible way. Further, the fixed number of states does not account for arbitrary large, extensible vocabularies.

While these previous random text models mainly aim at producing a Zipfian distribution of word frequencies, the new model presented here also takes the properties of neighbouring co-occurrence into account and introduces the notion of sentences in random text. After having pointed out the deficiencies of related models, a generation process is provided that takes neither the Zipf distribution on word frequencies nor the Small World property of the neighbouring co-occurrence graph as a constraint. Nevertheless, these distributions emerge in the process. The distributions obtained with the random generation model are compared to a sample of natural language data, showing high agreement also on word length and sentence length.

3.3.2 Desiderata for Random Text Models

The construction of the novel random text generation model proceeds according to the following guidelines [see 145]:

- *simplicity*: a generation model should reach its goal using the simplest mechanisms possible but results should still comply to characteristics of real language
- *plausibility*: without claiming that the model is an exhaustive description of what makes human brains generate and evolve language, there should be at least a possibility that similar mechanisms can operate in human brains.
- *emergence*: rather than constraining the model with the characteristics we would like to see in the generated stream, these features should emerge in the process.

The model described here is basically composed of two parts that will be described separately: a *word generator* that produces random words composed of letters, and a *sentence generator* that composes random sentences of words. Both parts use an internal graph structure, where traces of previously generated words and sentences are memorised. A key notion is the strategy of following 'beaten tracks': letters, words and sequences of words that have been generated before are more likely to be generated again in the future. But before laying out both generators in detail, ways of testing agreement of a random text model with natural language text are introduced.

3.3.3 Testing Properties of Word Streams

All previous approaches aimed at reproducing a Zipfian distribution on word frequency, which is a criterion that certainly has to be fulfilled. But there are more characteristics that should be obeyed to make a random text more language-like than previous models:

- *Lexical spectrum*: the smoothness or step-wise shape of the rank-frequency distribution affects the lexical spectrum, which is the probability distribution on word frequency. In natural language texts, this distribution follows a power-law with an exponent close to 2 [cf. 96].
- *Distribution of word length*: according to Sigurd et al. [223], the distribution of word frequencies by length follows a variant of the gamma distribution, which is given by $F(x) = x^a \times b^x$.
- *Distribution of sentence length*: the random text's sentence length distribution should resemble natural language. In [223], the same variant of the gamma distribution as for word length is fitted to sentence length.
- *Significant neighbour-based co-occurrence*: at random generation without word order awareness, the number of word pairs that are significantly co-occurring in neighbouring positions should be very low. In text generation, the number of significant co-occurrences in natural language as well as the graph structure of the neighbour-based co-occurrence graph should be reproduced.

The latter characteristic refers to the distribution of words *in sequence*. As written language is rather an artefact of the most recent millennia than a realistic sample of everyday language, the beginning of the spoken language section of the BNC is used for comparison to test the model. For simplicity, all letters are capitalised and special characters are removed, such that merely the 26 letters of the English alphabet are contained in the sample. Being aware of the fact that letter transcription is in itself an artefact of written language, this is chosen as a good-enough approximation. The sample contains 1 million words in 125,395 sentences with an average length of 7.975 words, which are composed of 3.502 letters in average.

3.3.4 Word Generator

The word generator emits sequences of letters, which are generated randomly in the following way: the word generator starts with a directed graph G_l with all N letters it is allowed to choose from as vertices. Initially, all vertices are connected to themselves with weight 1. When generating a word, the generator chooses a letter x according to the letter's probability $P(letter = x)$, which is computed as the normalised weight sum of outgoing edges:

$$P(letter = x) = \frac{weightsum(x)}{\sum_{i \in G_l} weightsum(i)} \tag{3.2}$$

where $weightsum(x) = \sum_{v \in V(G_l)} ew(x,v)$.

Then the word generator proceeds with the next position. At every position, the word ends with a probability $w \in (0,1)$. Otherwise, the next letter is generated according to the letter production probability given above. For every letter bigram, the weight of the directed edge between the preceding and current letter in the letter graph is increased by one. This results in self-reinforcement of letter probabilities: the more often a letter is generated, the higher its weight sum will be in subsequent steps, leading to an increased generation probability. Figure 3.17 shows how a word generator with three letters A, B, C changes its weights during the generation of the words AA, BCB and ABC. The word generating function is given in Algorithms 1 and 2: here, `concatenate(.,.)` returns a string concatenation of its two string arguments, `random-number` returns a random number between 0 and 1. In the implementation, the Mersenne Twister [160] is used for random number generation.

Algorithm 1 initialise-letter-graph G_l

insert $X_1, X_2, ..X_N$ vertices in $V(G_l)$
insert edges $e(X_i, X_j)$ for $i, j \in \{1,2,..N\}$ into $E(G_l)$ with weight 0
insert edges $e(X_i, X_i)$ for $i \in \{1,2,..N\}$ into $E(G_l)$ with weight 1

Algorithm 2 generate-word

word=(empty string)
previous-letter=(empty string)
repeat
  next-letter=pick letter randomly according to $P(letter)$
  word=concatenate(word,next-letter)
  if not previous-letter equals (empty string) **then**
    increase weight of e(previous-letter,next-letter) in G_l by 1
  end if
  previous-letter=next-letter
until random-number < w
return word

The word generator alone does produce a smooth Zipfian distribution on word frequencies and a lexical spectrum following a power-law. Figure 3.18 shows the frequency distribution and the lexical spectrum of 1 million words as generated by the word generator with $w = 0.3$ on 26 letters in comparison to a Mandelbrot generator with the same parameters. As depicted in Figure 3.18, the word generator fulfils the requirements on Zipf's law and the lexical spectrum, yielding a Zipfian exponent of around 1 and a power-law exponent of around 2 in the lexical spectrum, both matching the values as observed previously in natural language in e.g. [256] and [96]. In contrast to this, the Mandelbrot model has a step-wise rank-frequency distribution and a distorted lexical spectrum. The word generator itself is therefore already an improvement over previous models as it produces a smooth Zipfian dis-

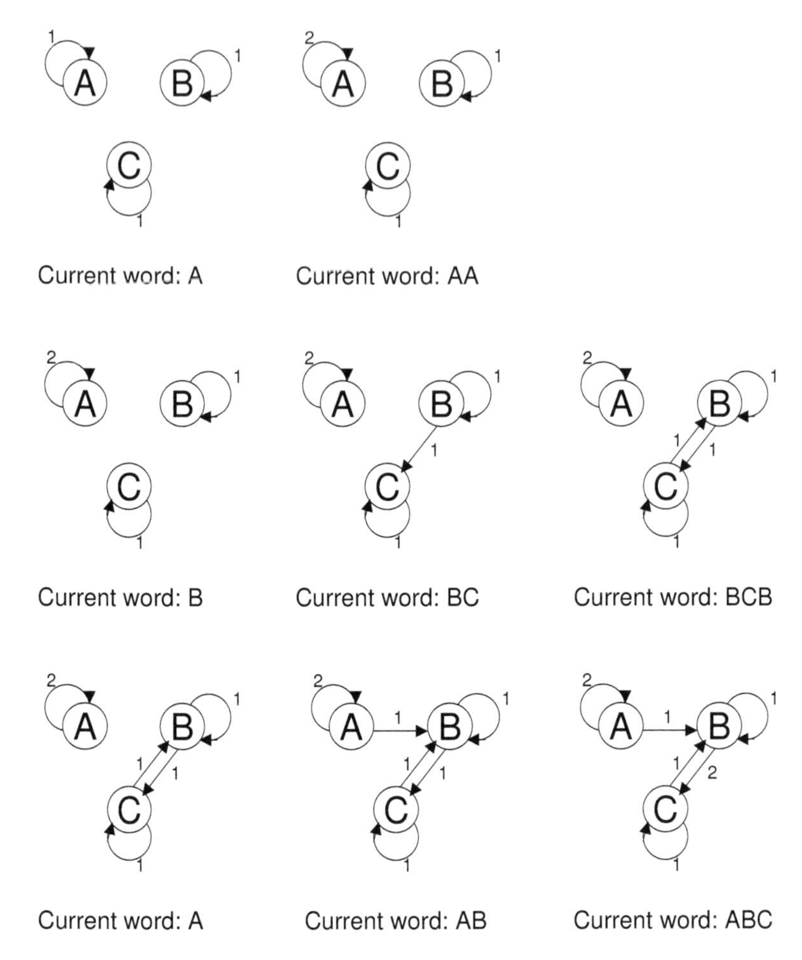

Fig. 3.17 Letter graph of the word generator during the generation of three words. The small numbers next to edges are edge weights. The probability for the letters for the next step are $P(letter = A) = 0.375$, $P(letter = B) = 0.375$ and $P(letter = C) = 0.25$

tribution and a lexical spectrum following a power-law, incorporating a notion of letters with emergent letter probabilities. But to comply to the other requirements, the process has to be extended by a sentence generator.

The reader might notice that a similar behaviour could be reached by just setting the probability of generating a letter according to its relative frequency in previously generated words. The graph seems an unnecessary complication for that reason. But retaining the letter graph with directed edges gives rise to a more plausible morphological production in future extensions of this model, probably in a way similar to the sentence generator.

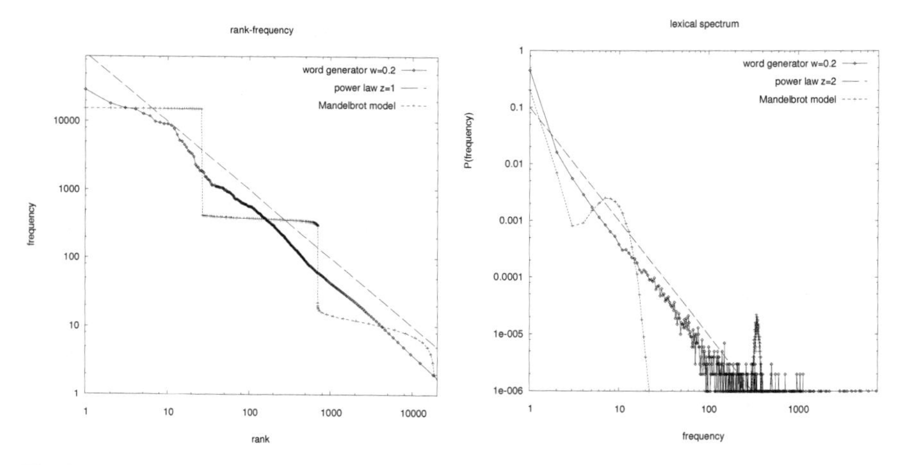

Fig. 3.18 Rank-frequency distribution and lexical spectrum for the word generator in comparison to the Mandelbrot model

3.3.5 Sentence Generator

The sentence generator model retains a directed graph G_w. Here, vertices correspond to words and edge weights correspond to the number of times two words were generated in a sequence. The word graph is initialised with a begin-of-sentence (BOS) and an end-of-sentence (EOS) symbol, with an edge of weight 1 from BOS to EOS. When generating a sentence, a random walk starts at the BOS vertex. With a probability $(1-s)$, an existing edge is followed from the current vertex to the next vertex according to its weight: the probability of choosing endpoint X from the endpoints of all outgoing edges from the current vertex C is given by

$$P(word = X|C) = \frac{ew(C,X)}{\sum_{N \in neigh(C)} ew(C,N)} \tag{3.3}$$

Otherwise, with probability $s \in (0,1)$, a new word is generated by the word generator, and a next word is chosen from the word graph in proportion to its weighted indegree: the probability of choosing an existing vertex E as successor of a newly generated word N is given by

$$P_{succ}(word = E) = \frac{indgw(E)}{\sum_{v \in V} indgw(v)} \tag{3.4}$$

with

$$indgw(X) = \sum_{v \in V} ew(v,X) \tag{3.5}$$

For each generated sequence of two words, the weight of the directed edge between them is increased by 1. Figure 3.19 shows the word graph for generating in sequence: (empty sentence), AA, AA BC, AA, (empty sentence), AA CA BC AA, AA CA CA BC. The sentence generating function is given in Algorithms 3 and 4.

Algorithm 3 initialise-word-graph G_w

```
insert BOS vertex into V(G_w)
insert EOS vertex into V(G_w)
insert e(BOS,EOS) into E(G_w) with weight 1
```

Algorithm 4 generate-sentence

```
current-vertex=BOS
repeat
  r=random-number
  if r < s then
    new-word=generate-word
    if new-word ∈ V(G_w) then
      increase weight of e(current-vertex,new-word) by 1
    else
      insert new-word into V(G_w)
      insert e(current-vertex,new-word) into E(G_w) with weight 1
    end if
    sentence=concatenate(sentence, new-word)
    successor=pick word randomly according to P_succ(word)
    increase weight of e(new-word,successor) by 1
    sentence=concatenate(sentence,successor)
    current-vertex=successor
  else
    next-word=pick word randomly according to P(word|current-vertex)
    sentence=concatenate(sentence,next-word)
    increase weight of e(current-vertex,next-word) by 1
    current-vertex=next-word
  end if
until current-vertex equals EOS
return sentence
```

During the generation process, the word graph grows and contains the full vocabulary used so far for generating in every time step. It is guaranteed that a random walk starting from BOS will finally reach the EOS vertex. It can be expected that sentence length will slowly increase during the course of generation as the word graph grows and the random walk has more possibilities before finally arriving at the EOS vertex. A similar effect, however, can be observed in human language learning, where children start with one- and two-word sentences and finally produce sentences of more than 1000 words in case of growing up to philosophers [see 168]. The sentence length is influenced by both parameters of the model: the word end probability w in the word generator and the new word probability s in the sentence generator. By feeding the word transitions back into the generating model, a reinforcement of previously generated sequences is reached. Figure 3.20 illustrates the sentence length growth for various parameter settings of w and s.

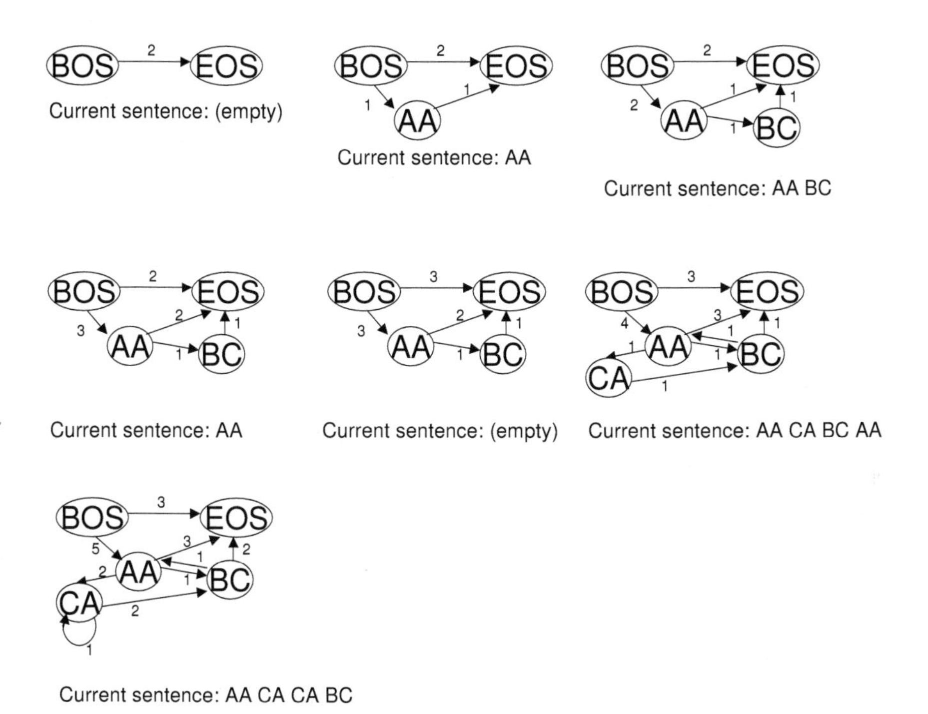

Fig. 3.19 Word graph of the sentence generator model. Notice that in the last step, the second CA was generated as a new word from the word generator, and so was the last AA in the previous sentence. The generation of empty sentences (omitted in output) happens frequently

When setting the new word probability s to 0 and not updating weights in G_w, the model is equivalent to a stationary Markov process with a horizon of 1. However, in this case, an initialisation with a fixed vocabulary and given transition probabilities would be necessary, which would reduce to a standard word bigram model. Both word and sentence generator can be viewed as weighted finite automata [cf. 10] with self training.

After having defined a random text generation model, the subsequent section is devoted to testing it according to the criteria given above.

3.3.6 Measuring Agreement with Natural Language

To measure agreement with the BNC sample, random text was generated with the sentence generator with $w = 0.4$ and $N = 26$ to match the English average word

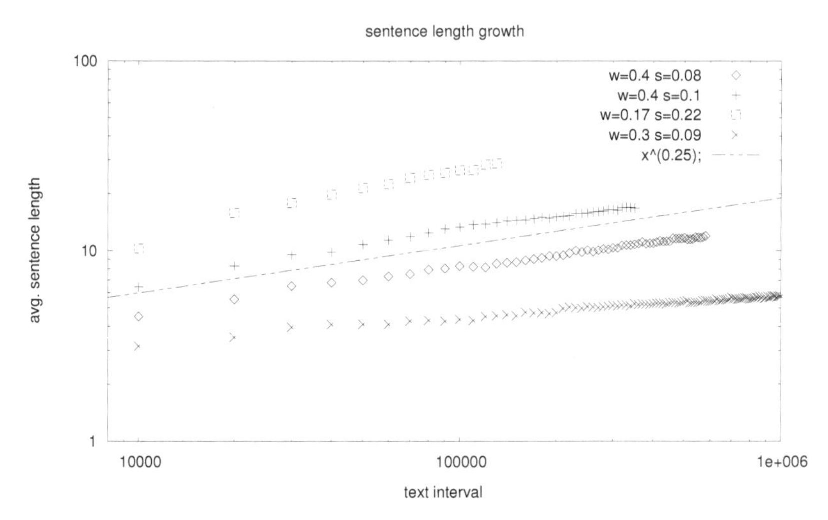

Fig. 3.20 Sentence length growth, plotted in average sentence length per intervals of 10,000 sentences. The straight lines in the log-log plot indicate polynomial growth

length and setting $s = 0.08$. The first 50,000 sentences were skipped to reach a relatively stable sentence length throughout the sample. To make the samples comparable, 1 million words totalling 12,552 sentences with an average sentence length of 7.967 are used.

3.3.6.1 Rank-Frequency

The comparison of the rank-frequency distributions for English and the sentence generator is depicted in Figure 3.21.

Both curves follow a power-law with γ close to 1.5, in both cases the curve is flatter for high frequency words as observed by Mandelbrot [157]. This effect could not be observed to this extent for the word generator alone (cf. Figure 3.18).

3.3.6.2 Word Length

While the word length in letters is the same in both samples, the sentence generator produced more words of length 1, more words of length>10 and less words of medium length. The deviation in single letter words can be attributed to the writing system being a transcription of phonemes — and few phonemes being expressed with only one letter. However, the slight quantitative differences do not oppose the similar distribution of word length in both samples, which is reflected in a curve of similar shape in Figure 3.22 and fits well the gamma distribution variant of Sigurd et al. [223].

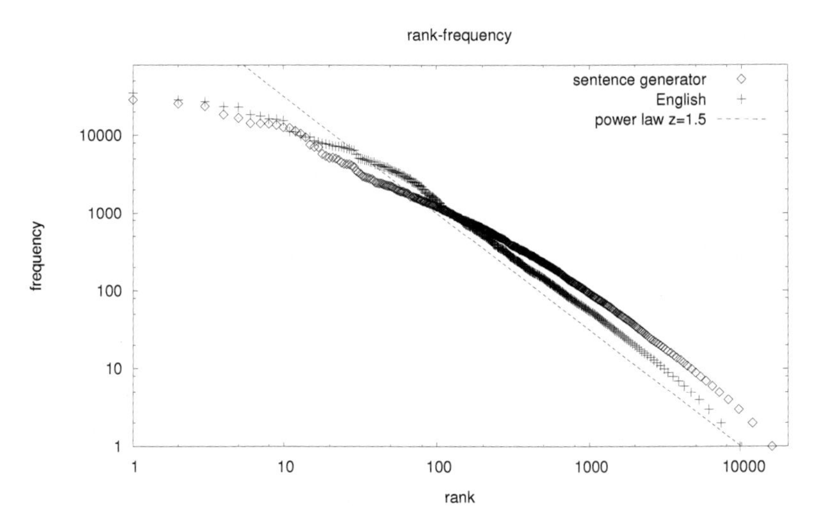

Fig. 3.21 Rank-frequency plot for English and the sentence generator

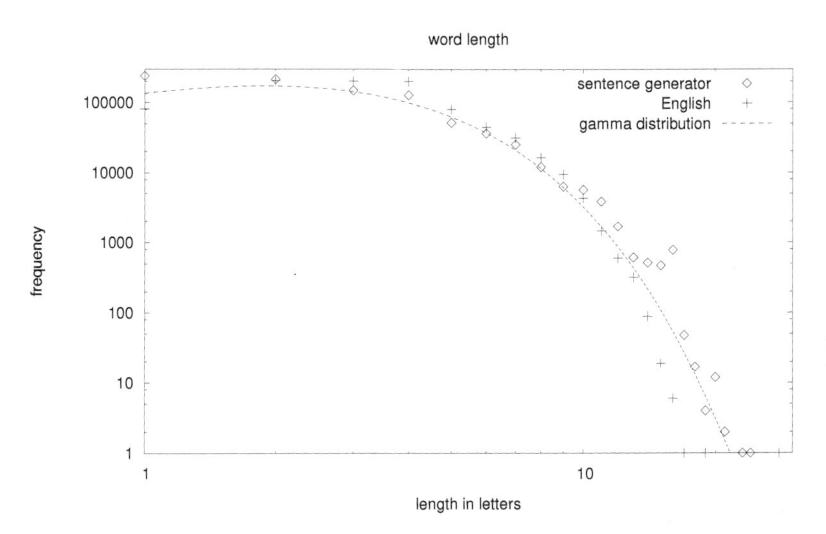

Fig. 3.22 Comparison of word length distributions. The dotted line is the function as introduced in [223] and given by $f(x) \sim x^{1.5} \times 0.45^x$

3.3.6.3 Sentence Length

The comparison of sentence length distribution shows again a high capability of the sentence generator to model the distribution of the English sample. As can be seen in Figure 3.23, the sentence generator produces less sentences of length>25 but does not show much differences otherwise.

In the English sample, there are surprisingly many two-word sentences. Inspection of the English sample revealed that most of the surprisingly long sentences are prayers or song lyrics, which have been marked as a single sentence.

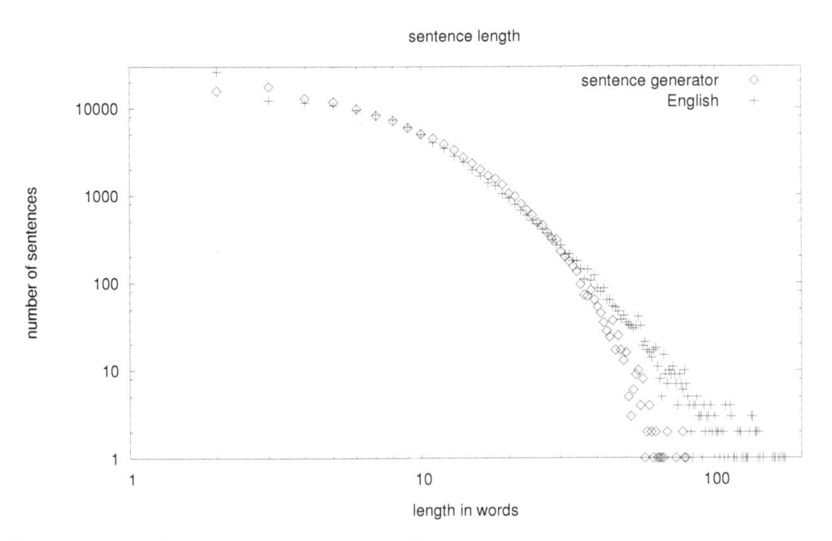

Fig. 3.23 Comparison of sentence length distributions

3.3.6.4 Neighbour-based Co-occurrence

Next, the structure of the significant neighbour-based co-occurrence graphs is examined. The significant neighbour-based co-occurrence graph contains all words as vertices that have at least one co-occurrence to another word exceeding a certain significance threshold. The edges are undirected and weighted by significance. Recall that the neighbour-based co-occurrence graph is scale-free and the Small World property holds, cf. Section 3.2.1.

For comparing the sentence generator sample to the English sample, log likelihood statistics on neighbouring words are computed with $t = 2$ and $s = 3.84$. For both graphs, the number of vertices, the average shortest path length, the average degree, the clustering coefficient and the degree distribution are given in Figure 3.24. Further, the characteristics of a comparable random graph as generated by the

ER-model and the graph obtained from the 1 million words generated by the word generator model are shown.

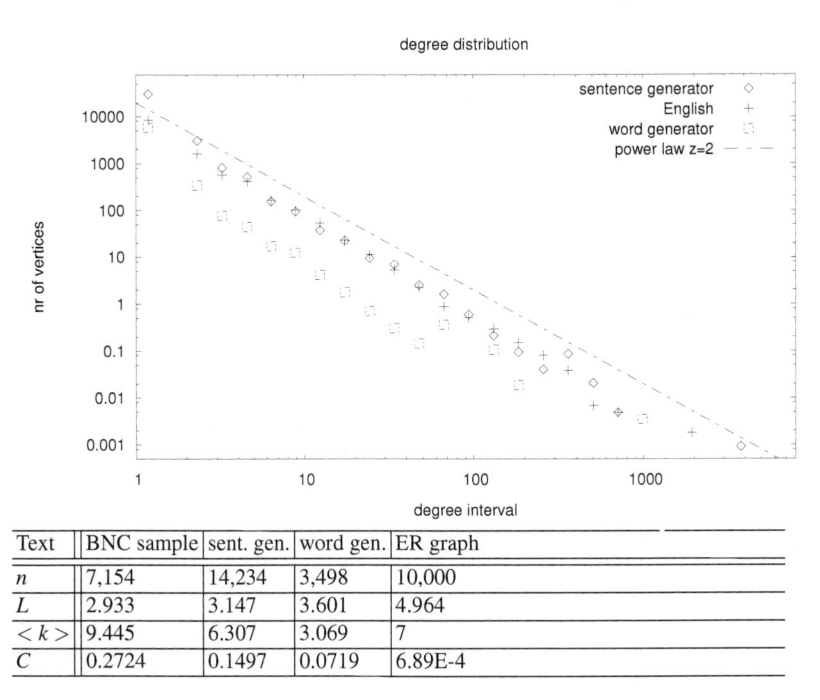

Text	BNC sample	sent. gen.	word gen.	ER graph
n	7,154	14,234	3,498	10,000
L	2.933	3.147	3.601	4.964
$<k>$	9.445	6.307	3.069	7
C	0.2724	0.1497	0.0719	6.89E-4

Fig. 3.24 Characteristics of the significant neighbour-based co-occurrence graphs of English and the generated samples of word and sentence generator

From the comparison with the random graph it is clear that both neighbour-based graphs exhibit the Small World property as their clustering coefficient is much higher than in the random graph while the average shortest path lengths are comparable. In quantity, the graph obtained from the generated sample has about twice as many vertices as the English sample, but its clustering coefficient is about half as high. This complies to the steeper rank-frequency distribution of the English sample (see Figure 3.21), which is, however, much steeper than the average exponent found in natural language. The degree distributions clearly match a power-law with $\gamma = 2$. The word generator data produced a number of significant co-occurrences that lies in the range of what can be expected from the 5% error of the statistical test. The degree distribution plot appears shifted downwards about one decade, clearly not producing the distribution of words in sequence of natural language. Considering the analysis of the significant neighbour-based co-occurrence graph, the claim is supported that the sentence generator model reproduces the characteristics of word sequences in natural language on the basis of bigrams.

3.3.6.5 Sentence-based Co-occurrence

It is not surprising that the sentence generator produces high agreement with natural language regarding neighbour-based significant word co-occurrences, since the random text model incorporates a notion of local word order. But also when measuring sentence-based word co-occurrence, the agreement is high, as Figure 3.25 shows: the sentence-based word co-occurrence graph ($t = 2$ and $s = 6.63$) for the generated text and the BNC sample show a very similar degree distribution, whereas the word generator produces less significant co-occurrences, although the difference is not as large as for word bigrams.

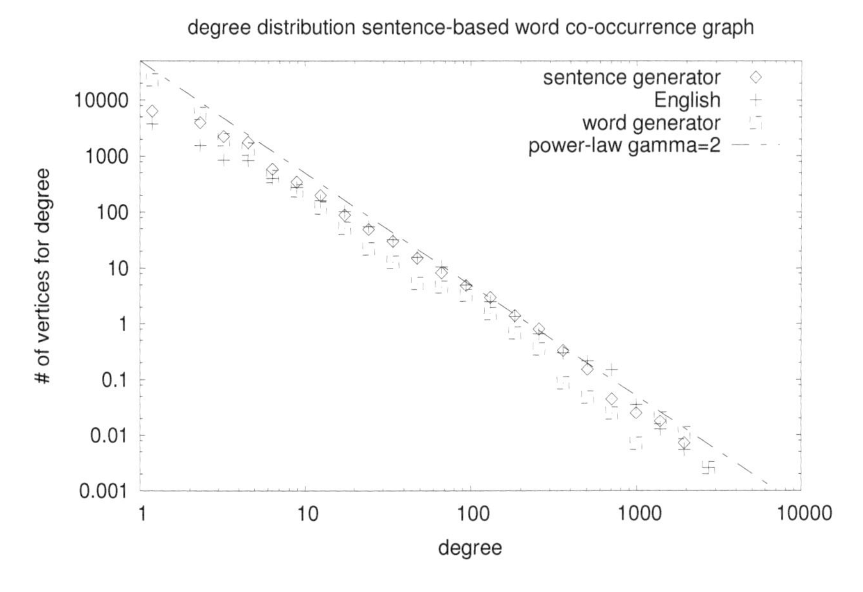

Fig. 3.25 Degree distributions of the significant sentence-based co-occurrence graphs of English, sentence generator and word generator. The differences are similar to measurements on neighbour-based co-occurrences, but not as distinctive

3.3.7 Summary for the Generation Model

A random text generation model was introduced that fits well with natural language with respect to frequency distribution, word length, sentence length and neighbouring co-occurrence. The model was not constrained by any a priori distribution — the characteristics emerged from a two-level process involving one parameter for the word generator and one parameter for the sentence generator. This is the first emergent random text generator that models sentence boundaries beyond inserting a special blank character at random: rather, sentences are modelled as a path between sentence beginnings and sentence ends, which imposes restrictions on the words at sentence beginnings and endings. Considering its simplicity, a plausible model for the emergence of large-scale characteristics of language is proposed without assuming a grammar or semantics. After all, the model produces gibberish — but gibberish that is well distributed.

The studies of Miller [177] rendered Zipf's law un-interesting for linguistics, as it is a mere artefact of language — as it emerges when putting a monkey in front of a typewriter — rather than playing an important role in its production. From what we learned from examining degree distributions of co-occurrence graphs originating from the word generator, it even seems that the power-law slope of $\gamma \approx 2$ is already caused by this simple model, although the quantity of significant co-occurrences differs from real language in the absence of word order awareness. The power-law in the degree distribution of the co-occurrence graph seems to be the immediate consequence of the power-law in the rank-frequency distribution: the higher the frequency of a word is, the more other words it co-occurs with.

The new model introduced here does not only explain Zipf's law, but many other characteristics of language, which are obtained with a monkey that follows beaten tracks. These additional characteristics can be thought of as artefacts as well, but I strongly believe that the study of random text models can provide insights in the process that lead to the origin and the evolution of human languages.

For further work, an obvious step is to improve the word generator so that it produces phonologically plausible sequences of letters and to intertwine both generators for the emergence of word categories. At this, letters could be replaced by phonemes, allowing only possible transitions as studied in the field of phonotactics [see 147]. This would not mean an a-priori initialisation of the model but a necessary adaptation to the human phonetic apparatus. To get the generator to produce something like word classes that could be found by e.g. the method proposed in Chapter 6, the word following a newly produced word could be not chosen by in-degree, but from the set of words following the followers of the word before the new word. Another suggestion is to introduce more levels with similar generators, e.g. for morphemes, phrases and paragraphs. For evaluation of these additional generators, however, much more precise quantitative measures would be needed.

Furthermore, it is desirable to embed the random generator in models of communication where speakers parameterise language generation of hearers and to examine, which structures are evolutionary stable [see 130]. This would shed light on the interactions between different levels of human communication. In this context, it would be desirable to ground the utterances to real-world situations, as carried out in the talking heads experiments of Steels [228] and to simulate on a realistic social network, as proposed in [164].

Chapter 4
Graph Clustering

Abstract This chapter is devoted to discovering structure in graphs by grouping their vertices together in meaningful clusters. In Section 4.1, clustering in its broadness is briefly reviewed in general. The discipline of graph clustering is embedded into the broad field of clustering and discussed in detail and a variety of graph clustering algorithms are examined in terms of mechanism, algorithmic complexity and adequacy for scale-free Small World graphs. Taking their virtues and drawbacks into consideration, an efficient graph partitioning algorithm called *Chinese Whispers* is developed in Section 4.2. It is time-linear in the number of edges, finds the number of clusters automatically and does not impose relative size restrictions on clusters, which is adequate for graphs from language data. Several extensions and parametrisations of the method are discussed. This algorithm will be used throughout later chapters to solve several NLP tasks.

4.1 Review on Graph Clustering

This section provides a review on graph clustering methods and their properties. After a introduction to clustering in general and graph clustering in particular, the distinction between spectral and non-spectral methods is drawn and several algorithms are described in more detail. Readers familiar with the graph clustering literature might skip to Section 4.2, where a new graph clustering algorithm will be defined that addresses the shortcomings of existing algorithms for language graphs.

4.1.1 Introduction to Clustering

Clustering is the process of grouping objects in a way that similar objects are found within the same group. Informally, the similarity between objects within a group should be higher than the similarity between objects of different groups, which are called clusters. Clustering is *data exploration*, in contrast to classification, which is the learning of a classifier that assigns objects to a given set of classes. Clustering is also known as unsupervised machine learning.

Among researchers, there is no consensus of what exactly constitutes a cluster. In [165], different measures are given that determine the quality of clusterings, yet the optimisation of a clustering with respect to a particular measure often results in a performance drop in other measures. Therefore the question cannot be how to produce an optimal clustering by itself, but to produce the best clustering for a given task, where the grouping of objects is used to achieve an improvement over not using a grouping or over an inferior way of grouping.

To be able to set up a clustering method, a measure is needed that defines a similarity (or dissimilarity) value for pairs of objects, represented by features. Independent of the clustering method, the similarity measure has a large influence on the result. Depending on the representation of objects (i.e. what kind of features are used), a plethora of similarity measures are possible and have to be chosen in order to meet the invariants of objects. For example, when clustering handwritten digits for optical character recognition, the shape of digits is more discriminative than their horizontal or vertical displacement or their size. Traditionally, objects are given as data points in a high-dimensional space, where the dimensions represent features that can be determined by examining the object. For the handwritten digits, every pixel could play the role of a feature, with the value set to the colour intensity of the pixel. The object's data point in the *feature space* is often referred to as *feature vector*.

Here, only the most basic differences of clustering methods are displayed rather than elaborated, to merely locate the discipline of graph clustering in the overall field of clustering. For recent extensive overviews of clustering methods, the reader is referred to [93; 20].

Different methods of clustering can be classified into hierarchical methods and partitionings. Hierarchical methods arrange the objects in nested clusters, following divisive (top-down) or agglomerative (bottom-up) strategies: In divisive clustering, all objects are regarded as being contained in one large cluster in the beginning. Clusters are iteratively split until either only single objects are found in the clusters of the lowest hierarchy level or a stopping criterion is met. Agglomerative methods proceed the opposite way: after initialisation, every object constitutes its own cluster. Iteratively, clusters are joined to form larger clusters of clusters until all objects are in one big cluster or the process stops because none of the intermediate clusters are similar enough to be merged. For the notion of similarity of clusters, again, various strategies are possible to compute cluster similarity based on the similarity of the objects contained in the clusters, e.g. single-link, complete-link and average-link cluster similarity, which are defined as the highest, lowest or the average similarity of pairs of objects from different clusters. Commonly used is the distance of centroids: a cluster centroid is an artificial data point whose coordinates are computed by averaging over the coordinates of all data points in the cluster. The benefit of hierarchical clustering methods, namely the arrangement of objects into nested clusters that are related to each other, is also its largest drawback: in applications that need a sensible number of non-nested clusters (such as the handwritten digits, which require 10 parts of the data), the question of where to place the vertical cut in the cluster hierarchy is non-trivial. Figure 4.1 shows the visualisation of a hierarchical clustering of points with similarity based on their distance in their 2-dimensional space, called dendrogram. The dotted lines are possible vertical cuts to turn the hierarchy into a flat partition.

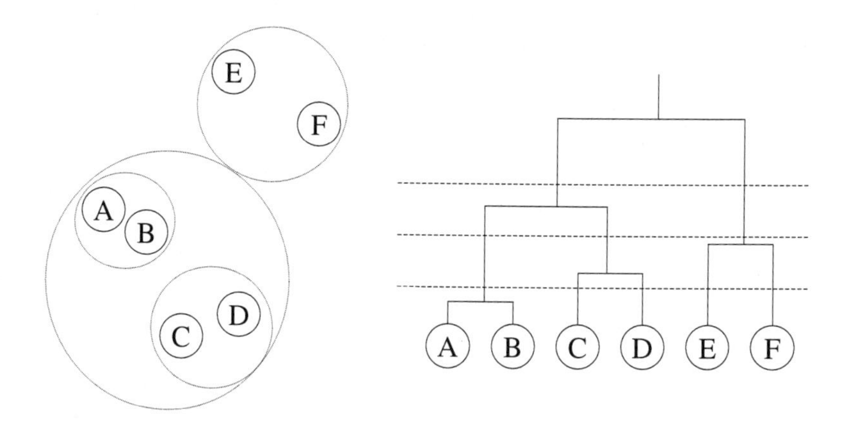

Fig. 4.1 Example of hierarchical clustering. Left: data points, right: dendrogram and possible vertical cuts for partitions $\{\{A,B,C,D\},\{E,F\}\}$, $\{\{A,B\},\{C,D\},\{E,F\}\}$ and $\{\{A,B\},\{C\},\{D\},\{E\},\{F\}\}$. Dotted circles display the nested clusters in the data; for similarity, the inverse distance between centres of these circles was used

The other option is to directly partition the objects into a set of disjoint sets of objects such that each object is member of exactly one cluster. This makes the vertical cut in the cluster hierarchy obsolete. The most prominent representative of partitioning algorithms is k-means [155]. The k-means algorithm retains k cluster centroids which are initialised randomly in the vector space spanned by the set of features. Each data point is assigned to the centroid to which it is most similar. Then, the cluster centroids are recomputed by averaging the feature vectors of the data points assigned to them. This is done iteratively until no changes can be observed in subsequent steps. K-means is very efficient, yet has two disadvantages: the number of clusters k has to be set beforehand and outcomes of different runs might differ due to the random initialisation of centroids: the algorithm gets stuck in local optima — a property that is shared by all randomised approaches.

The bulk of the literature on clustering describes methods operating on data points in an n-dimensional space, and a metric is used to compute similarity values. This is due to the nature of the data, which is naturally represented as feature vectors in many settings. An alternative, which will be used throughout this work, is the graph representation. Here, objects are represented as vertices in an undirected graph, and edge weights between these vertices reflect similarity.

Traditionally, the vector space model [208], originating from the field of information retrieval [see 13, for a survey], has been widely used in NLP and examined in great detail in [207, inter al.]. In Schütze's Word Space [219] and related works, an entity (e.g. a document) is represented as a high-dimensional feature vector, where features are usually words appearing in the document. Notice that it is straightforward to compute the graph representation from the vector space representation by constructing a graph with one vertex per object and computing pair-wise similarities on the vector space that form the weights of edges between the respective vertices. This conversion is lossy, as it is not possible to restore the original feature vectors from the similarity graph. As clustering algorithms normally operate on similarities rather than on the single feature vectors, however, this loss does not render graph-based methods less powerful.

Variants of this conversion are possible: an edge between two vertices can be drawn only if the weight exceeds a certain threshold. Notice that some entities in the vector representation might not end up in the graph in this case. Another variation could be adding only the N most similar adjacent vertices for each vertex.

There are fundamental differences between vector space and graph representation. First of all, vector spaces have dimensions, which are spanned by the features respectively the vector entries. There is nothing corresponding to a dimension in graphs — they encode similarity directly instead of assigning a location in a high-dimensional space. For this reason, the vector space notion of a centroid as the average vector of a set of vectors does not have an equivalent in graphs. A consequence of thresholding the minimum weight in the conversion is the possible unrelatedness of individual entities in the graph: whereas a suitable similarity function assigns non-zero values to all vector pairs, and thus grades the similarity scores, there is a possibly large fraction of vertex pairs in the graph with zero similarity to all other vertices, which have to be excluded from the clustering.

The conversion algorithm outlined above has a quadratic time-complexity in the number of entities, as each possible pair of entities has to be considered. This is unwanted, as we shall see that further processing steps can have much lower time-complexities which would turn this conversion into a bottleneck. A heuristic to optimise the conversion is to index the term-document matrix by feature and to compare only pairs of entities that show to be amongst the top N entities which have this feature set to a non-zero value.

When transforming data that is represented in a vector space, the question arises why to do so. Dependent on the task to be solved, this may or may not be advantageous.

In some cases, feature vector-based techniques are not applicable, as the data is already given in a graph representation. This is often the case in the field of natural language processing, and this is why this work concentrates mainly on graph methods. For a comprehensive textbook on graph methods for natural language processing and information retrieval, see [174].

In the next section, two different paradigms of graph clustering are described, one operating directly on graphs, the other operating on the spectrum of its adjacency matrix.

4.1.2 Spectral vs. Non-spectral Graph Partitioning

Spectral graph clustering is based on spectral graph theory, see [63; 74] for thorough overviews of this field. Spectral methods make use of invariants of the adjacency matrix (especially its Laplacian[1]), such as characteristic polynomials, eigenvalues and eigenvectors. This is closely related to dimensionality reduction techniques like singular value decomposition (SVD) and latent semantic analysis (LSA, [79]). E.g. the algorithm of Shi and Malik [222] splits the vertices in two parts based on the second-smallest eigenvalue of the Laplacian. A related setup is used by Meilă and Shi [166], where the eigenvectors for the largest k eigenvalues of a adjacency matrix invariant are used as initial centroids for k-means, which of course requires a vector space representation, which is derived from the adjacency matrix. For spectral clustering, the eigenvectors corresponding to the largest k eigenvalues define the clusters, which means that spectral clustering always needs the number of clusters k as an input parameter. In general, spectral methods rely on various matrix operations, most of which cause the entire algorithm to be slower than time-quadratic[2] in the number of vertices, making spectral techniques only applicable to graphs of at most several thousand vertices. As in NLP, often millions of vertices are needed to represent e.g. different words in a corpus, and dimensionality reduction techniques have been shown to be generally inadequate for the power-law distributions found in natural language [cf. 150; 232], the focus of this work will be on non-spectral methods in the remainder. Non-spectral methods operate directly on the graph or its adjacency matrix without computing the invariants as described above. Methods are often inspired by 'classical' graph theory, which is outlined in e.g. [39] or [108]. In the next subsection, a number of non-spectral graph clustering algorithms are summarised.

4.1.3 Graph Clustering Algorithms

Given a graph, the question arises how it should be split into two or more parts. Intuitively, a cluster in a graph is connected with many internal edges, and there are only few edges between clusters. In graph theory, several measures have been proposed to determine how vertices and groups of vertices relate to each other.

---

[1] the Laplacian matrix is given by $L = D_G - A_G$, where D_G is the degree matrix of graph G that contains the degree of vertices in the main diagonal.

[2] iterative improvements in matrix multiplication complexity suggest that its time complexity is quadratic, although the fastest known algorithms are slower, see [68].

4.1.3.1 Measures of Cluster Quality

Šíma and Schaeffer [224] summarise several locally computable fitness functions, which are used for measuring the quality of a cluster within a graph.

The *conductance* of a cut S, $\phi(S)$, is defined as the size of the cut, divided by the minimum of the sums of degree in the parts:

$$\phi_G(S) = \frac{c_G(S)}{\min(\sum_{a\in S}(k_S(a), \sum_{b\in V/S}(k_{V/S}(b)))} \quad (4.1)$$

For clustering, the conductance of a cut should be minimised. The conductance of a weighted graph is also called *normalised cut*, as used by Shi and Malik [222].

The local density of a subset $\emptyset \neq S \subset V$ in a graph G is the ratio of the number of edges in the subgraph $G(S)$ induced by S over the number of edges in a clique of $|S|$ vertices:

$$ldensity_G(S) = \frac{2|E(S)|}{(|S| \times (|S|-1))}. \quad (4.2)$$

The local density measures the ratio between actual and possible edges in a set of vertices. For clustering, the local density of parts should be maximised.

The relative density of a cut S is defined as

$$rdensity_G(S) = \frac{|E(S)|}{(|E(S)| + c_G(S))} \quad (4.3)$$

and measures the amount edges that are within the parts versus the amount that are in the cut. Cluster quality is higher for larger relative density.

Single cluster editing of a subset $S \subset V$ counts the number of edge operations (additions of edges and deletions of edges) needed to transform S into an isolated clique and can be computed as

$$sincluedit_G(S) = \binom{|S|}{2} - |E(S)| + c_G(S). \quad (4.4)$$

Single cluster editing should be minimised for a good clustering.

Many graph clustering algorithms search for clusters that directly optimise these measures [see e.g. 44; 131; 169; 212].

In [224], proofs are given that optimising each single of the four measures is NP-complete, i.e. the number of computations needed for the optimisation is exponential in the number of vertices, making the optimisation intractable: no efficient solutions exist for large numbers of vertices[3]. Therefore, algorithms that optimise these measures are theoretically justified, but are not applicable for large graphs because their run-times are too long to be of practical use. Other NP-complete measures and approximations thereof can be found in [22]. In [134], cases where the

[3] if $P \neq NP$ holds for the classes P and NP in computational complexity, which is unproven, yet widely accepted.

optimisation of graph measures even leads to undesired results are discussed and exemplified.

4.1.3.2 Desiderata on Graph Clustering for Natural Language

Graph partitioning has been of practical use mainly in the field of very large scale integration (VLSI), being a part of hardware design for processors. A major current question in VLSI is to distribute digital logic on several chips. Atomic circuits can be modelled as vertices in a graph, where edges represent dependencies between circuits. To arrive at a performant architecture, the number of dependencies between different chips should be minimised, because the transfer between chips is much slower than within chips. An additional constraint in VLSI is that the parts are of roughly equal size to implement the overall circuit on several equisized chips. See [100] for a survey on graph partitioning in VLSI and [125] for an evaluation of bipartitioning techniques. Unfortunately, methods developed for VLSI are only applicable with restrictions to language data, as the constraint of a fixed number of equisized parts does not reflect the natural structure of language: as indicated in Chapter 3, the distribution of cluster sizes in language data can be expected to follow a power-law, which implies a highly skewed distribution on natural cluster sizes rather than clusters of similar scale. As an example, the number of word types in syntactic word classes differ considerably: whereas open word classes like nouns, adjectives and verbs contain a large number of members, closed word classes like determiners, prepositions and pronouns only posses a few. A clustering algorithm on word classes that aims at equisized clusters would split open word classes into several clusters while crudely lumping closed word classes together. Another difference between applications like VLSI and natural language is that in contrast to an engineer's task of distributing a current network over a fixed number of chips, in natural language a sensible choice of the number of clusters is often not known beforehand. Think of the number of topics in a text collection: without reading the collection, a user will not be able to specify their number, but clustering into topics is exactly what a user wants in order to avoid reading the whole collection. Other examples include the number of different languages in a random internet crawl, which will be discussed in depth in Chapter 5, and the number of word meanings in word sense induction, see Chapter 7.

Facing the NP-completeness of the measures above, a creator of an efficient graph clustering algorithm is left with two choices: either to optimise a measure that is not NP-complete, or to give computationally low-cost heuristics that only approximate one of the measures above. Both strategies can be observed in a selection of graph clustering algorithms that will be presented in the following.

4.1.3.3 Graph Clustering Review

The most widely known polynomial-time graph measure is *mincut* (minimal cut): a graph is (iteratively) split in a top-down manner into two parts in each step, such that the size of the cut is minimised. Finding a minimal cut is equivalent to finding the maximal flow between two vertices in a current network as stated in the max-flow min-cut theorem [102]. The global minimal cut is achieved by minimising over max-flows between all pairs of vertices. The (1+ε) approximate mincut problem can be solved for weighted, undirected graphs in $O(|E|+|V|^{4/3}\varepsilon^{-16/3})$ as proposed by Christiano et al. [62] and has therefore viable run-time properties for graphs with several thousand vertices. For larger graphs, Wu and Leahy [249] describe a method to transform the graph into a cut-equivalent tree that allows even faster processing times. Another alternative of scaling is the use of parallelisation [see 136]. Mincut is used for building trees on graphs by Flake et al. [101] for web community identification. A problem of mincut, however, is that several minimum cuts can exist, of which many can be counter-intuitive. Figure 4.2 illustrates two possible outcomes of mincut on an unweighted graph with cut size of 4.

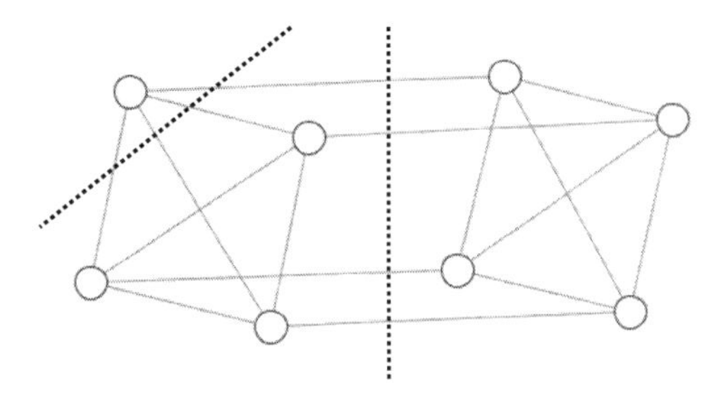

Fig. 4.2 Two possible minimal cuts for an unweighted graph of 8 vertices. Cutting only one vertex off is counter-intuitive and does not optimise any of measures given above

The counter-intuitiveness of the 'bad' minimal cut in Figure 4.2 originates in the fact that one of the 4-cliques is cut apart. To overcome this flaw by balancing cluster sizes, *ratio cuts* were used by Wei and Cheng [246] for VLSI and normalised cuts by Shi and Malik [222] for image segmentation. In [222], a spectral approximation for minimising conductance only results in tolerable run-times for pictures of some 10,000 pixels when taking properties of image segmentation into account: e.g. that the graphs are only locally connected — again a property that does not hold for graphs in natural language, opposing the shortest average path lengths in word graphs, see Chapter 3. When using mincut to iteratively split a graph in a hierarchical setting, the number of clusters or a maximum cut value has to be defined, making the methods subject to individual tuning for each dataset.

More appropriate for the clustering of Small World graphs seems Small World Clustering [161]. Here, the weighted sum of average shortest path length and clustering coefficient are to be maximised by removing and adding edges until a local maximum is found. Non-connected components of the resulting graph constitute the clusters. The author tests the approach on a small word co-occurrence network. Since the computation of both measures is computationally expensive, a more efficient heuristic would be needed in order to make the procedure feasible for large graphs.

Much faster and also employing the Small World property is the HyperLex algorithm of Veronis [240]. Here, the following steps are repeated on an undirected graph, edge-weighted by dissimilarity of vertex pairs: the largest hub is identified as being a root vertex, and all of its neighbours are deleted from the graph, until no more hub vertex with a sufficient degree and local clustering coefficient can be found. These root vertices form the first level of a minimum spanning tree (that is, the tree containing all vertices of the graph with the minimum total edge weight), which contains the clusters of vertices in its first level subtrees. The minimum spanning tree can be computed in $O(|E|log|E|)$, which makes the overall procedure very efficient. HyperLex was developed and evaluated for word sense induction, cf. Chapter 7. The various parameters for determining whether a vertex is counted as hub, however, are subject to application-dependent fine-tuning.

An interesting point is raised by Ertöz et al. [91] in the context of graph-based document clustering: a general problem with clustering algorithms is that regions with a higher density, i.e. many data points in a local neighbourhood in the vector space or highly connected subgraphs in the graph model, tend to be clustered together earlier than regions of low density. It might be the case that in hierarchical settings, two dense clusters that are close to each other are agglomerated while sparse regions are still split into small pieces, resulting in an unbalanced clustering. Figure 4.3 shows an illustrating example.

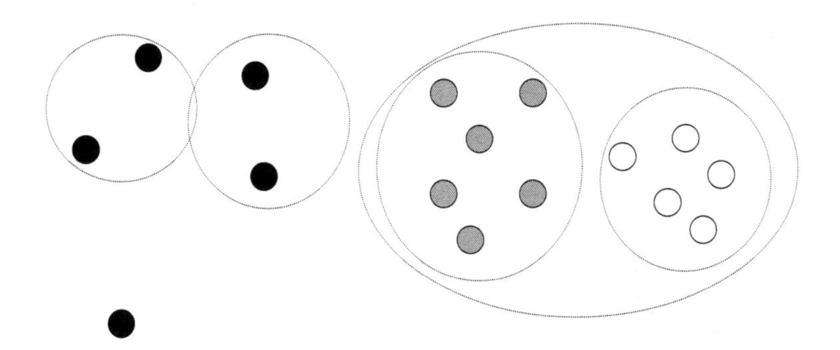

Fig. 4.3 Dense and sparse regions: the dense clusters are joined before the sparse region is clustered. In hierarchical settings, this situation leads to no vertical cut that would produce the intuitively expected clusters. In partitioning the same problem arises with either too many small clusters in the sparse region or too large clusters in the dense region. Shades indicate the desired partition; the dotted circles show two levels of hierarchical clustering

To overcome this problem, Ertöz et al. [91] introduce self scaling in the graph model: for each vertex, only the N most similar (highest-weighted) neighbours are considered. A new graph is computed from this top-N graph by counting the number of different paths of length 2 between any pair of vertices and assigning the result as edge weights in the new graph. The more common neighbours two vertices had in the top-N graph, the higher is their resulting weight in the new graph, cf. the higher order transformation described in Section 3.2.2. As the operation of taking the top N neighbours is not attributed with a weight or similarity threshold, this operation adapts to the density differences in the original graph. Further, Ertöz et al. [91] describe the necessity for various similarity thresholds: for clustering, a higher threshold is used to determine the partition. As thresholding might result in a portion of vertices without edges, these are attributed to their nearest cluster in a subsequent step that uses a lower threshold. In this way, vertices with low edge weights do not contribute to the cluster structure but are nevertheless part of the final clustering.

Elaborating on the bipartite structure of many datasets (e.g. terms and documents in information retrieval), Zha et al. [253] apply methods designed for general graphs to the bipartite setting and define bipartite cuts and bipartite normalised cuts. The divisive binary clustering algorithm described there is time-linear in the number of vertices, which is achieved by an approximation using singular value decomposition (which has to be computed beforehand). Here, again, the number of relevant dimensions determines the number of clusters and has to be provided as a parameter.

Another way of looking at graph clustering is based on the notion of a random walk on graphs. A random walk is the formalisation of the idea of taking successive steps at random along possible paths. In graphs, a random walk is a path through vertices of a graph along the edges; in weighted graphs, the probability of walking along an edge is proportional to its weight. Random walks on graphs can be viewed as Markov chains: the next step is only dependent on the vertex (state) where the random walker is positioned at a certain time step, and time invariance is given by the fact that the probabilities of successor vertices (edge weights) are invariant of specific time steps.

Random walks have been applied to rank vertices according to their prominence in PageRank [48] for internet search engines, further for text summarisation in [175] and [90] and other applications [174]. Prominence ranking using random walks is based on the intuition that the more important a vertex is, the more often it is visited by a long random walk.

Using random walks in clustering is based on another idea: random walks starting at some point in a cluster should be ending up in the same cluster after a couple of steps more likely than ending up in another cluster because of a higher inter-cluster than intra-cluster connectivity that constitutes a desired clustering. This idea is employed in Markov Chain Clustering (MCL, [82]). Since MCL has been used in NLP for word sense induction by Widdows and Dorow [247] and is related to the clustering algorithm that will be introduced in the next section, it is explained in more detail.

MCL is the parallel simulation of all possible random walks up to a finite length on a graph G. MCL simulates flow on a graph by repeatedly updating transition probabilities between all vertices, eventually converging to a transition matrix after k steps that can be interpreted as a clustering of G. This is achieved by alternating an expansion step with an inflation step. The expansion step is a matrix multiplication of the adjacency matrix A_G with the current transition matrix. The inflation step is a column-wise non-linear operator that increases the contrast between small and large transition probabilities and normalises the column-wise sums to 1. The k matrix multiplications of the expansion step of MCL lead to a time-complexity worse than quadratic. The outcome of the algorithm is deterministic and depends sensitively on four parameters that have to be fine-tuned by the user: a loop weight for allowing the random walk to stay at a vertex, two inflation constants for influencing the cluster granularity, and an initial prefix length for manipulating the input data. The number of clusters is found by the method itself.

It has been observed by van Dongen [82] that only a few early iterations operate on dense matrices — when using a strong inflation operator, matrices in the later steps tend to be sparse. The author further discusses pruning schemes that only keep a few of the largest entries per column dependent on an additional pruning parameter, leading to drastic optimisation possibilities. But the most aggressive sort of pruning is not considered: only keeping one single largest entry. This and other optimisations will be undertaken in the next section, at the price of ending up with a randomised and non-deterministic, yet more efficient algorithm.

4.2 Chinese Whispers Graph Clustering

In this section, the Chinese Whispers (CW) graph clustering algorithm is described. CW was first published in [34] and later more formally described in [27]. Both publications form the basis for this section. Additionally, extensions of CW are discussed. The development of this algorithm is very central to this work: the algorithm is used throughout Chapters 5, 6 and 7 to realise Structure Discovery procedures. However, it should be stressed that any other suitable clustering algorithm fulfills the goals of Structure Discovery, and that the SD paradigm is not tied in any way to graph clustering in general and to Chinese Whispers in particular.

As data sets in NLP are usually large, there is a strong need for efficient methods, i.e. of low computational complexities. In the previous section, several graph clustering algorithms were examined on their utility for these datasets. It turned out that algorithms that exactly optimise graph-theoretical measures are mostly intractable, and that most approximations either aim at producing equisized clusters or take the number of desired clusters as input parameter. Now, a very efficient graph clustering algorithm is introduced that is capable of partitioning very large graphs in comparatively short time. This is realised by merely taking local properties of vertices into account, in contrast to optimising global measures like the minimal cut. The number of parts emerges in the process and does not have to be defined by the user. While this is reached by a non-deterministic and formally not converging algorithm, the method is applicable efficiently to datasets that are prohibitively large for most other methods.

4.2.1 Chinese Whispers Algorithm

CW is a very basic — yet effective — algorithm to partition the vertices of weighted, undirected graphs. The name is motivated by the eponymous children's game (in American English known as 'telephone'), where children whisper words to each other[4]. While the game's goal is to arrive at some funny derivative of the original message by passing it through several noisy channels, the CW algorithm aims at finding groups of vertices that broadcast the same message to their neighbours. It can be viewed as a simulation of an agent-based social network; for an overview of this field, see [12]. The algorithm is outlined as follows:

Algorithm 5 Standard Chinese Whispers CW(graph $G(V,E)$)

```
for all v_i ∈ V do
    class(v_i) = i
end for
for it=1 to number-of-iterations do
    for all v ∈ V, randomised order do
        class(v)=predominant class in neigh(v)
    end for
end for
return partition P induced by class labels
```

Intuitively, the algorithm works in a bottom-up fashion: first, all vertices get different classes. Then the vertices are processed in random order for a small number of iterations and inherit the predominant class in the local neighbourhood. This is the class with the maximum sum of edge weights to the current vertex. In case of multiple predominant classes, one is chosen randomly. This behaviour on predominant class ties and the random processing order of vertices within an iteration render the algorithm non-deterministic.

Regions of the same class stabilise during the iteration and grow until they reach the border of a stable region of another class. Notice that classes are updated continuously: a vertex can obtain classes from the neighbourhood that were introduced there in the same iteration. The fractions of a class a in the neighbourhood of a vertex v is computed as

$$fraction(a,v) = \frac{\sum_{w \in neigh(v), class(w)=a} ew(v,w)}{\sum_{w \in neigh(v)} ew(v,w)}, \tag{4.5}$$

the predominant class a in the neighbourhood of v is given by

$$\arg\max_{a} fraction(a,v). \tag{4.6}$$

[4] the name was given by Vincenzo Moscati in a personal conversation.

For each class label in the neighbourhood, the sum of the weights of the edges to the vertex in question is taken as score for ranking. Figure 4.4 illustrates the change of a class label.

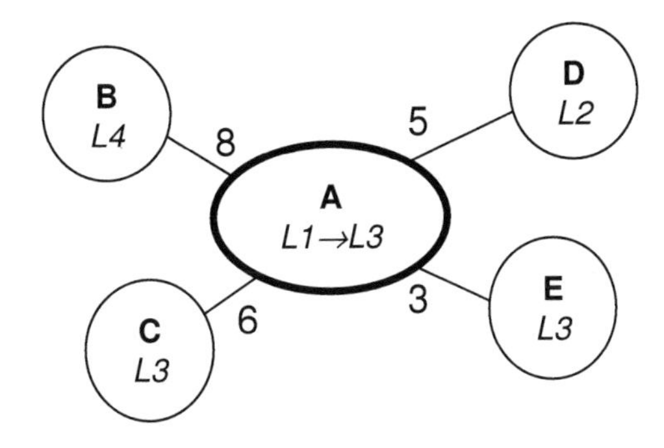

Fig. 4.4 The class label of vertex A changes from L1 to L3 due to the following scores in the neighbourhood: L3:9, L4:8 and L2:5

With increasing iterations, clusters become self-preserving: if a strongly connected cluster happens to be homogeneous with respect to class labels, it will never be infected by a few connections from other clusters.

Figure 4.5 illustrates how a small unweighted graph is clustered into two regions in two iterations. Different classes are symbolised by different shades of grey.

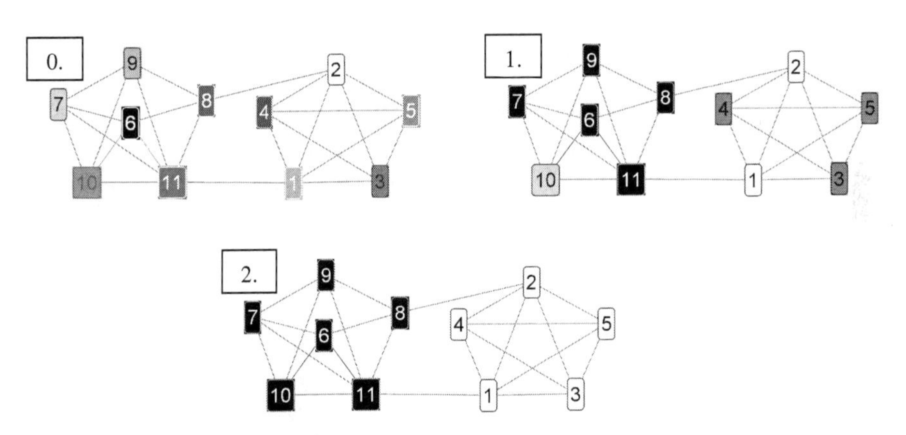

Fig. 4.5 Clustering an 11-vertices graph with CW in two iterations

The CW algorithm cannot cross component boundaries, because there are no edges between vertices belonging to different components along which class labels could be spread. Further, vertices that are not connected by any edge are discarded from the clustering process, which possibly leaves a portion of vertices unclustered. Formally, CW does not converge, as Figure 4.6 exemplifies: here, the middle vertex's neighbourhood consists of a tie that can be decided in assigning the class of the left or the class of the right vertices in any iteration all over again. Ties, however, do not play a major role in most weighted graphs.

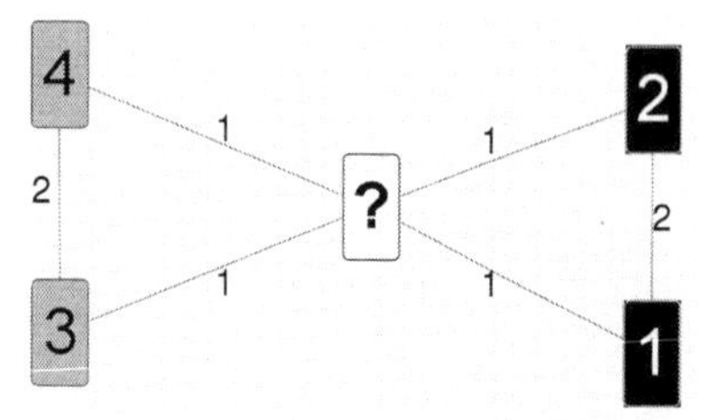

Fig. 4.6 The middle vertex gets assigned the grey or the black class. Small numbers denote edge weights

Apart from ties, classes usually stabilise after a handful of iterations. The number of iterations depends on the diameter of the graph: the larger the distance between two vertices is, the more iterations it takes to percolate information from one to another.

The result of CW is a hard partition of the given graph into a number of parts that emerges in the process — CW is parameter-free. It is possible to obtain a soft partition by assigning a class distribution to each vertex, based on the weighted distribution of (hard) classes in its neighbourhood in a final step.

In an unweighted setting, isolated n-cliques always get assigned the same class label. This is due to the reason that a vertex typing with two different class labels in a n-clique is not stable under CW iterations, as there are always more edges to the larger one of the parts than to vertices having the same label for sizes up to $\frac{n}{2}$. Recursively applying this argument leads to homogeneity of classes in n-cliques.

CW can be compared to the Potts Model [194] in statistical mechanics, which models phase transitions of spins in chrystalline lattices. The lattice is replaced by a graph in CW, where coupling strength is given by edge weights. As opposed to the approaches in statistical mechanics, CW does not attempt to optimally solve the model, but approximates a solution by aggressively propagating only a single label at a time.

Another algorithm that uses only local contexts for time-linear clustering is DBSCAN as described in [92], which depends on two input parameters (although the authors propose an interactive approach to determine them). DBSCAN is especially suited for graphs with a geometrical interpretation, i.e. the objects have coordinates in a multidimensional space. A quite similar algorithm to CW is MAJORCLUST [230], which is based on a comparable idea but converges slower, because vertex type changes do not take effect until the next iteration. The LPA algorithm of Raghavan et al. [199] is an independent later re-discovery of Chinese Whispers.

As CW can be viewed as a special and altered case of MCL (van Dongen [82], see Section 4.1.3), similarities and differences are examined. The main contribution in the complexity of MCL originates from the multiplication of non-sparse matrices, which is alleviated through pruning most row entries to zero. In CW, the same pruning is done but with the difference of keeping only one single non-zero entry per row, i.e. one class label per vertex.

One could try writing CW as a matrix process in the style of MCL [see 82]. Let `maxrow(.)` be an operator that operates row-wise on a matrix and sets all entries of a row to zero except the largest entry, which is set to 1. Then, the algorithm is denoted as simple as this:

Algorithm 6 matrix-CW(adjaceny matrix A_G)

$D_0 = I^n$
for t=1 to number-of-iterations **do**
  D_{t-1} = maxrow(D_{t-1})
  $D_t = D_{t-1}A_G$
end for

In Algorithm 6, t is time step, I^n is the identity matrix of size $n \times n$, A_G is the adjacency matrix of graph G and D_t is the matrix retaining the clustering at time step t.

By applying `maxrow(.)`, D_{t-1} has exactly n non-zero entries. This causes the time-complexity to be dependent on the number of edges, namely $O(|E|)$. In the worst case of a fully connected graph, this equals the time-complexity of MCL. However, as shall be clear from Section 2.2, Small World graphs are rather sparsely connected. A problem with the matrix CW process is that it does not necessarily converge to an iteration-invariant class matrix D_∞, but rather to a pair of oscillating class matrices. Figure 4.7 shows an example.

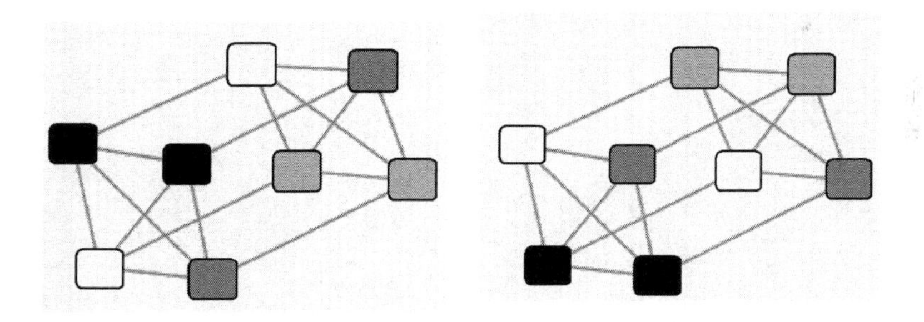

Fig. 4.7 Oscillating states in matrix CW for an unweighted graph

This is caused by the stepwise update of the class matrix: entries changed in the previous iteration do not take effect until the next iteration. As opposed to this, the standard CW as outlined in Algorithm 5 immediately updates D after the processing of each vertex. Apart from pruning with the `maxrow(.)` operator, this is the main difference between CW and MCL. Due to the immediate update mechanism, standard CW can not be expressed as a process involving matrix multiplications. To avoid these oscillations, alternative measures could be taken:

- *Random mutation*: with some probability, the maxrow operator places the 1 for an otherwise unused class.
- *Keep class*: with some probability, the row is copied from D_{t-1} to D_t.

While converging to the same limits, the continuous update strategy converges the fastest because prominent classes are spread much faster in early iterations. Random mutation showed to have negative effects when partitioning small graphs, as mutation weakens clusters in their formation phase, which then possibly gets overtaken by the labels of a strong neighbouring cluster.

4.2.2 Empirical Analysis

The analysis of the CW process is difficult due to its nonlinear and randomised nature. Its run-time complexity indicates that it cannot directly optimise most global graph cluster measures mentioned in Section 4.1.3. Therefore, experiments on synthetic graphs are performed to empirically arrive at an impression of the algorithm's abilities. All experiments were conducted with an implementation following Algorithm 5. Experiments with synthetic graphs are restricted to unweighted graphs, if not stated otherwise.

A cluster algorithm should keep dense regions together while cutting apart regions that are sparsely connected. The highest density is reached in fully connected sub-graphs of n vertices, a.k.a. n-cliques. An *n-bipartite-clique* is defined as a graph of two n-cliques, which are connected such that each vertex has exactly one edge going to the clique it does not belong to. Figures 4.7 and 4.8 are n-bipartite cliques for $n = 4, 10$.

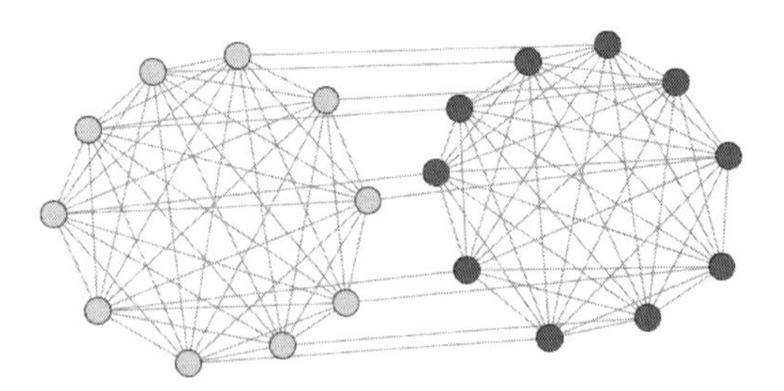

Fig. 4.8 The 10-bipartite clique

A graph clustering algorithm can be clearly expected to cut the two cliques apart. In the unweighted case, CW is left with two choices: producing two clusters of n-cliques or grouping all vertices into one cluster. This is largely dependent on the random choices in very early iterations — if the same class is assigned to several vertices in both cliques, it will finally cover the whole graph. Figure 4.9 illustrates on what rate this happens on n-bipartite-cliques for varying n.

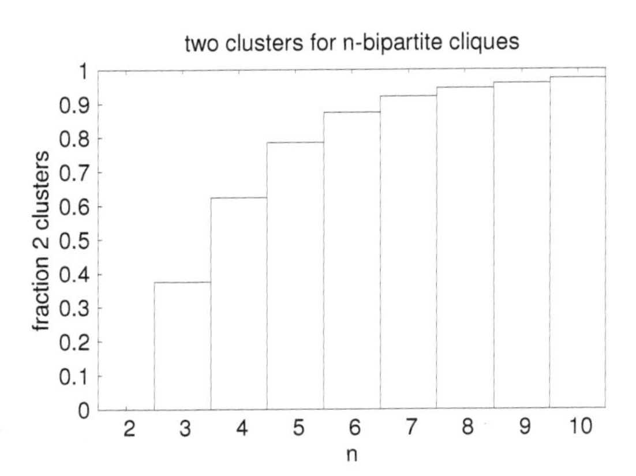

Fig. 4.9 Percentage of obtaining two clusters when applying CW on n-bipartite cliques. Data obtained from 10,000 runs/n

It is clearly a drawback that the outcome of CW is non-deterministic. Only two thirds of the experiments with 4-bipartite cliques resulted in separation. But the problem is only dramatic on small graphs and ceases to exist for larger graphs as demonstrated in Figure 4.9. These small clustering problems can also be solved exactly with other clustering algorithms — interesting in the context of NLP is the performance on larger graphs and graphs with the Small World property.

For the next experiment, it is assumed that natural systems can be characterised by Small World graphs. If two or more of those systems interfere, their graphs are joined by merging a part of their vertices, retaining their edges. A graph clustering algorithm should split the resulting graph in its original parts, at least if not too many vertices were merged. To measure CW's performance on SW-graph mixtures, graphs of various sizes were generated and merged in various extents in a pair-wise fashion. Now it is possible to measure the amount of cases where clustering with CW leads to the reconstruction of the original parts. Graphs were generated using the Steyvers-Tenenbaum model (see Section 2.2): the mean connectivity $2M$ was fixed to 20, the number of vertices n and the merge rate r, which is the fraction of vertices of the smaller graph that is merged with vertices of the larger graph, was varied.

Figure 4.10 summarises the results for equisized mixtures of 300, 3,000 and 30,000 vertices and mixtures of 300 with 30,000 vertices.

It is not surprising that separating the two parts is more difficult for higher r. Results are not very sensitive to size and size ratio, indicating that CW is able to identify clusters especially if they differ considerably in size. The separation quality depends on the merge rate, but also on the total number of common vertices.

As the algorithm formally does not converge, it is important to define a stopping criterion or to set the number of iterations. To show that only a few iterations are needed until almost-convergence, the normalised Mutual Information (nMI) between the clustering in the 50th iteration and the clusterings of earlier iterations was measured. The normalised Mutual Information between two partitions X and Y is defined as

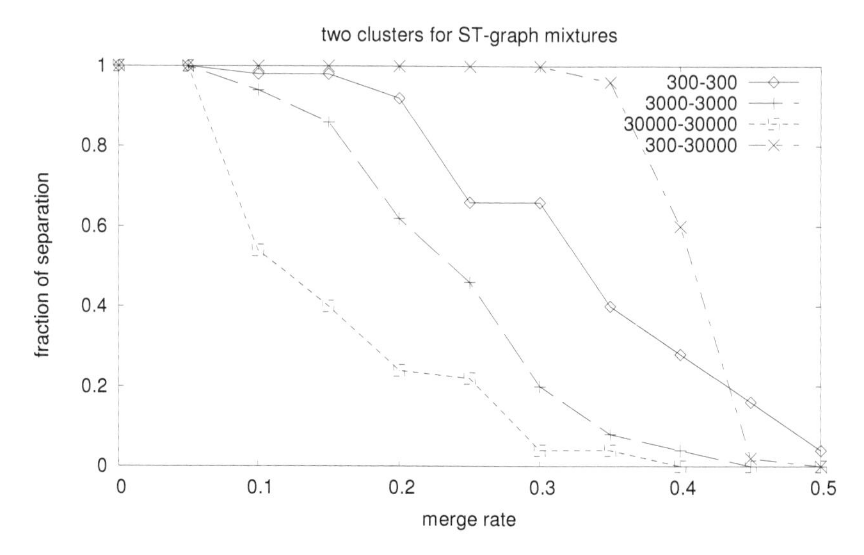

Fig. 4.10 Rate of obtaining two clusters for mixtures of SW-graphs dependent on merge rate r. Data obtained for 5 mixtures and 10 runs per mixture for each data point

$$nMI(X,Y) = \frac{(H(X)+H(Y)-H(X,Y))}{\max(H(X),H(Y))} \tag{4.7}$$

with entropy

$$H(X) = -\sum_{i=1..n} p(X_i)log(p(X_i)) \tag{4.8}$$

for $p(X_i)$ probability of choosing from part X_i when selecting a vertex from X randomly, $X_1..X_n$ forming a partition of X, and

$$H(X,Y) = -\sum_{x,y} p_{x,y}log(p_{x,y}) \tag{4.9}$$

joint entropy of X and Y. A value of 0 denotes independence, 1 corresponds to congruence of partitions X and Y.

This was conducted for two unweighted ST-graphs with 1,000 (1K) and 10,000 (10K) vertices, $M = 5$ and a weighted 7-lingual co-occurrence graph (cf. Section 5, as used in [34]) with 22,805 vertices and 232,875 edges. Table 4.1 indicates that for unweighted graphs, only small changes happen after 20-30 iterations. In iterations 40-50, the normalised MI-values do not improve any more. The weighted graph converges much faster due to fewer ties and reaches a stable plateau after only 6 iterations.

To summarise the empirical analysis, experiments with synthetic graphs showed that for small graphs, results can be inconclusive due to the randomised nature of CW. But while there exist a huge variety of clustering approaches that can deal well with small graphs, its power lies in its capability of handling very large graphs in

Table 4.1 Normalised Mutual Information values for three graphs and different iterations in %

Iter	1	2	3	5	10	20	30	40	49
1K	1	8	13	20	37	58	90	90	91
10K	6	27	46	64	79	90	93	95	96
7ling	29	66	90	97	99.5	99.5	99.5	99.5	99.5

edge-linear time. The applicability of CW rather reaches out in size regions, where other approaches' solutions are intractable. In Chapters 5, 6 and 7, CW, as well as some of its extensions that are discussed in the following, will be evaluated in the context of unsupervised NLP applications.

4.2.3 Weighting of Vertices

Until now, the update step of CW treats all neighbouring vertices equally. That means, the strength of neighbouring class labels only depends on the edge weights to the vertex to be updated, and not on other properties. If a graph possesses scale-free Small World structure, then it contains vertices with very high degrees (hubs), which try to propagate their class label to a large number of vertices. As the nature of those hubs implies that they have connections to many regions of the graph, this effect might be unwanted, as class labels can tunnel through them, probably resulting in several intuitive parts to obtain the same class — this especially happens if several hubs obtain the same class in early iterations. A possibility would be the deletion of hubs (see [9] for attacking the connectivity of graphs) — hence it is not clear how many of the vertices with very high degrees should be deleted. Another possibility is to downgrade the influence of hubs by assigning lower vertex weights vw to them. For the vertex weighted version, the fraction of a class a in the neighbourhood of a vertex v is rewritten as:

$$fraction(a,v) = \frac{\sum_{w \in neigh(v), class(w)=a} vw(w) \times ew(v,w)}{\sum_{w \in neigh(v)} vw(w) \cdot ew(v,w)}. \tag{4.10}$$

Note that the incorporation of vertex weights is a real extension, as the function $vw(.)$ is arbitrary and can lead to different effective weights of edges, dependent on which vertex is to be updated. For downgrading hubs, $vw(x)$ can be defined to be dependent on the degree of vertex x. Here, two variants are examined: $vw_{LIN}(x) = 1/k(x)$ and $vw_{LOG}(x) = 1/\log(1+k(x))$. Dependent on vertex weighting, different outcomes of update steps can be achieved, as Figure 4.11 illustrates.

Weighting down hubs should result in more and smaller clusters. To test this hypothesis, a ST-network (see Section 2.2) with 100,000 vertices and an average degree of $2M = 5$ was generated and clustered using the different vertex weighting schemes. Figure 4.12 shows the distribution of cluster sizes.

The data presented in Figure 4.12 supports the claim that a more rigidly down-weighting of hubs according to their degree results in more and smaller clusters.

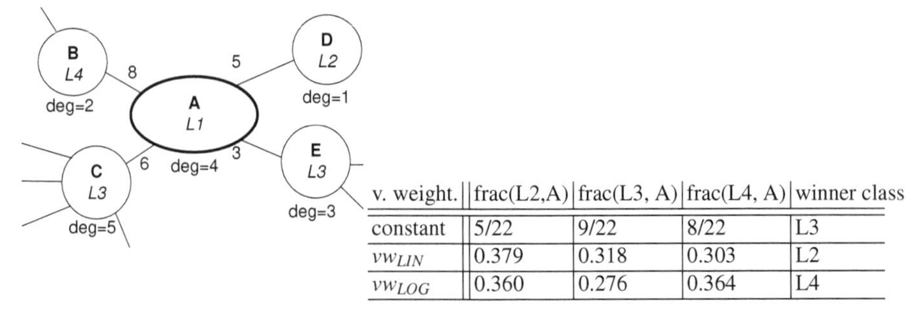

v. weight.	frac(L2,A)	frac(L3, A)	frac(L4, A)	winner class
constant	5/22	9/22	8/22	L3
vw_{LIN}	0.379	0.318	0.303	L2
vw_{LOG}	0.360	0.276	0.364	L4

Fig. 4.11 Effects of vertex weighting in the neighbourhood of vertex A. The table summarises the fractions of the different classes in the neighbourhood for different ways of vertex weighting. Standard CW is equivalent to constant vertex weighting

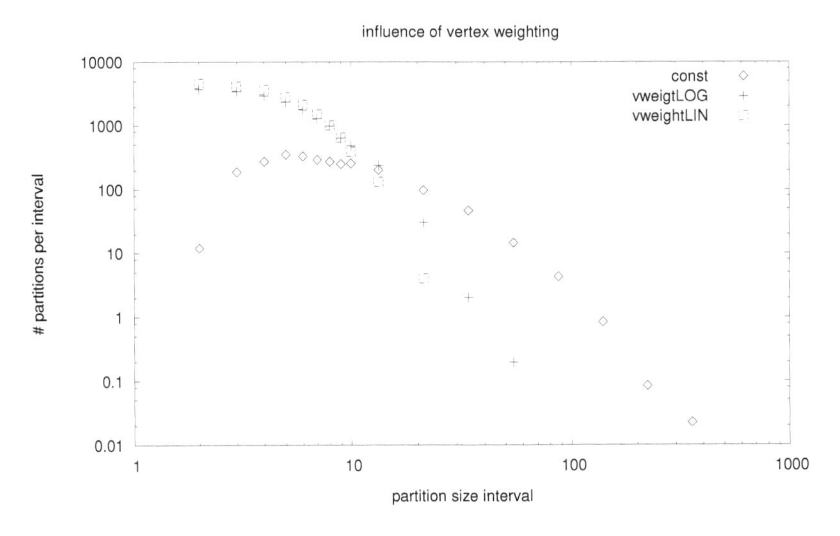

Fig. 4.12 Cluster size distribution for different vertex weighting schemes

Whereas in this experiment the largest cluster for the standard setting was 358 vertices, it was 46 vertices using vw_{LOG} and only 21 vertices in the case of vw_{LIN}. In summary, vertex weighting provides a means for influencing the granularity of the partition.

4.2.4 Approximating Deterministic Outcome

It might seem an undesired property of CW that different runs on the same graph can produce different outcomes, which is caused by the randomised nature of CW. This behaviour does not result in severe performance drops on real world data (see Chapters 5, 6 and 7), because there are several ways to split a graph in equally motivated clusters, which result in similar performance in applications. Still, a possibility of alleviating this problem is discussed here shortly. Consider the topped tetrahedron graph [cf. 82] in Figure 4.13.

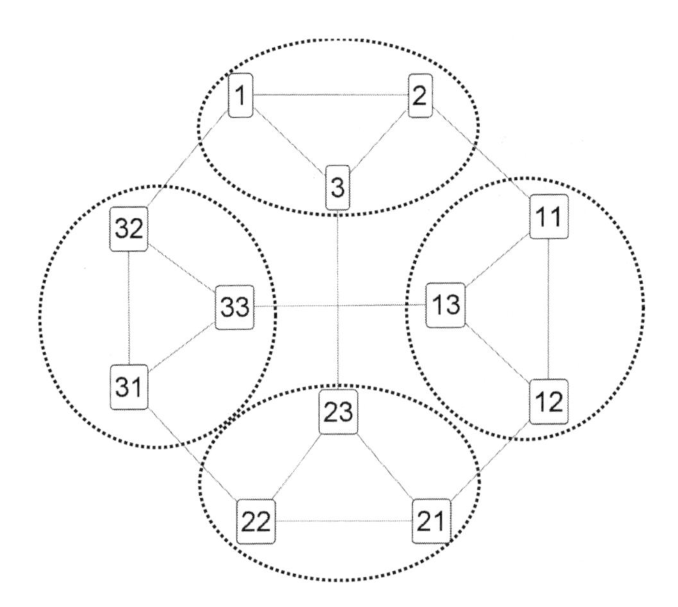

Fig. 4.13 Topped tetrahedron graph with plausible partition

A plausible partition would split this graph into four parts, consisting of one triangle each. As in the experiments with n-bipartite graphs, however, CW is able to find this partition, but can assign the same class to arbitrary combinations of these four parts, yielding 1, 2, 3 or 4 parts for this graph. Note that the four central nodes in the figure do not form a clique.

Deterministic Chinese Whispers combines several runs of standard CW in the following way: amongst all runs, the run having produced the most parts is chosen. This serves as a reference run all other partitions are aligned to. For each other run, a confusion matrix of class labels is computed: looking up the classes for every vertex in both partitions, the confusion matrix contains the number of times the two class labels of the reference run and the other run are found together. For the classes of the other run, a fuzzy mapping to the reference classes is then given by the vector of reference classes per other class. The final result is obtained by replacing the classes in all runs by their reference class vectors and summing up the vectors position-wise. Vertices with the same reference vectors belong to the same class. Figure 4.14 shows an example for three different partitions of the topped tetrahedron graph.

From the example it should be clear that the finest-grained partition possible is found by merging possibly overlapping partitions. The result partition does not

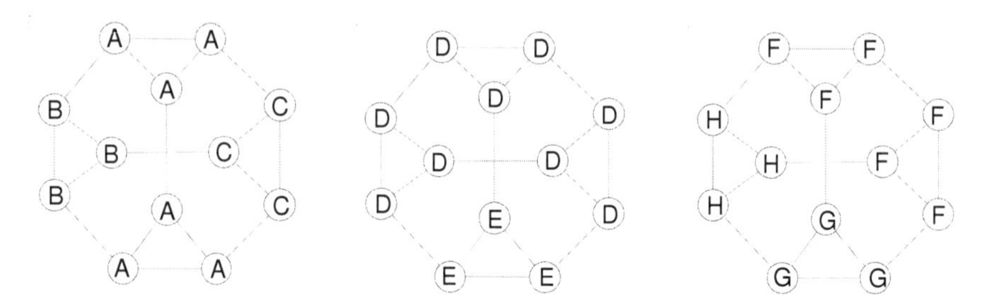

Fig. 4.14 Three partitions of the topped tetrahedron. The first run is selected for reference

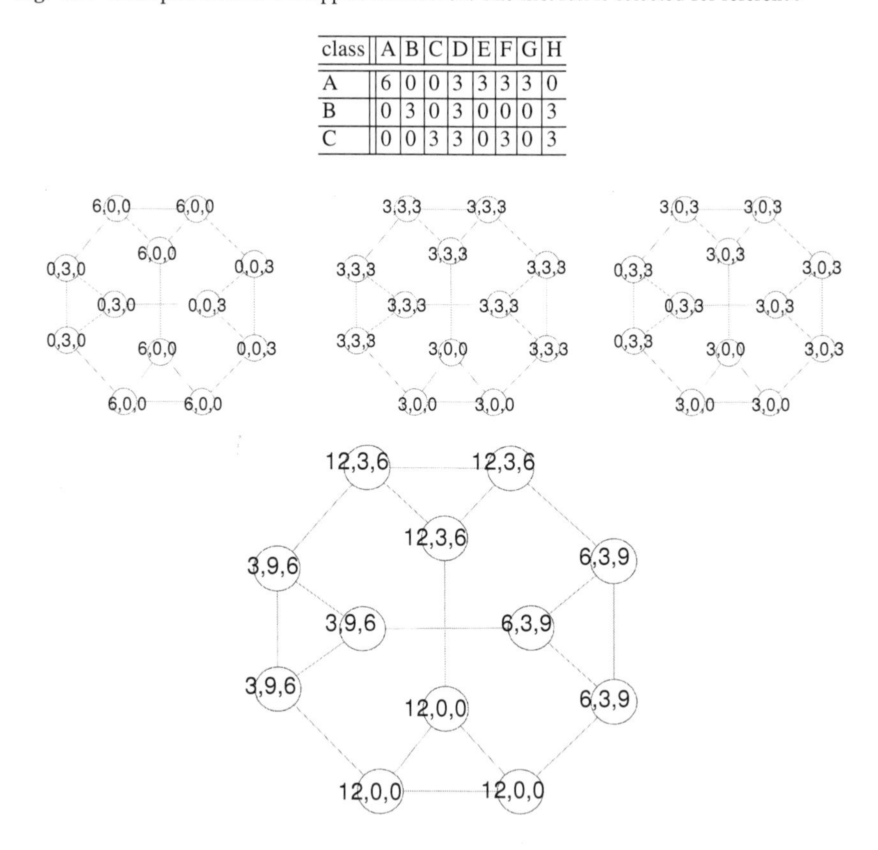

class	A	B	C	D	E	F	G	H
A	6	0	0	3	3	3	3	0
B	0	3	0	3	0	0	0	3
C	0	0	3	3	0	3	0	3

Fig. 4.15 Reference class vectors obtained from comparing run 1 with run 2 and run 1 with run 3, then result, which is equivalent to the plausible partition in Figure 4.13. Notice that no class occurs in more than one partition

have to be contained in any of the runs, and can be sometimes finer-grained than any partition of a single run: the graph in Figure 4.6 is split into three clusters in the result, while in both possible outcomes of single CW runs only two clusters are found. Picking the run with the maximum number of clusters as reference avoids picking a trivial partition, and aligning all other runs to it keeps the procedure linear in the number of vertices. Using enough runs, this procedure finds all pairs of vertices that can possibly end up in two different parts when applying CW. In this way, the outcome of CW can be made almost deterministic. The number of standard CW runs does not have to be fixed beforehand, but can be iteratively increased until no changes are observed in the result. The convergence test pins down to counting the number of different classes in the combined result, which increases monotonically until the maximum possible number is reached. As this operation involves n (number of vertices) operations, it does not increase the computational complexity of the overall process, yet larger graphs require a larger number of runs. Using this stopping criterion, however, does not give a guarantee that the maximal possible number of clusters is found, as the currently last run of standard CW does not increase the number of clusters if equivalent in result to a previous run. When taking several recent runs without changes into account, however, the possibility of premature convergence can be lowered.

4.2.5 Disambiguation of Vertices

Now, a graph transformation is described that performs the conversion of the original graph into a graph with disambiguated vertices. The transformation is motivated by the fact that ambiguity, i.e. the possibility to use the same symbol for several denotations, is omnipresent in natural language. The most widely used example for lexical ambiguity is "bank", which can be used in the sense of a financial institution or refer to a river bank. But also other levels of ambiguity, e.g. in compound noun splitting, part-of-speech, phoneme-to-text conversion and phrase attachment need to be resolved. In the graph framework, a vertex is said to be ambiguous, if it belongs to more than one group, therefore acting in different roles and having edges to vertices of all groups it is related to. This property of vertices is purely local and constitutes itself in the neighbourhood of the vertex, not being affected by other regions of the graph. Following the idea of Widdows and Dorow [247], the different roles or usages of a vertex v can be obtained by partitioning the subgraph induced by the neighbourhood of vertex v, $G(neigh(v))$. As $v \notin neigh(v)$, the subgraph may be formed of several components already. Partitioning $neigh(v)$ with CW results in $m \geq 1$ parts of $G(neigh(v))$. For the transformation, vertex v is replaced by m vertices $v@0, v@1, ...v@(m-1)$, and $v@i$ is connected to all vertices in cluster i, retaining edge weights. This is done iteratively for all vertices, and the results are combined in the disambiguated graph. Figure 4.16 shows an unweighted graph and its disambiguated graph. Note that all neighbourhoods of vertices apart from vertex 1 yield one cluster, whereas $neigh(1)$ is split into two clusters.

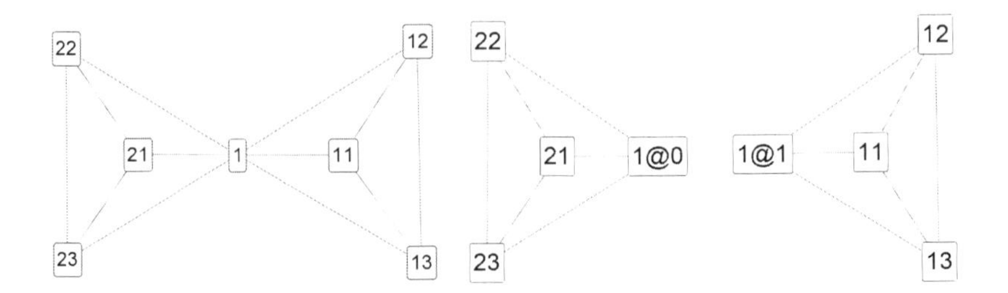

Fig. 4.16 Left: a graph. Right: its disambiguated graph

Whereas a CW-partition of the original graph can result in one cluster, the disambiguated version is always split into two clusters. Applying the disambiguation transformation to the n-bipartite cliques above results in splitting the graph in $n+2$ components, of which n reflect the connecting edges between the cliques, and the remaining two are formed by the cliques, see Figure 4.17. The small components reflecting the connecting edges can be pruned by not allowing singletons in the neighbourhood graphs.

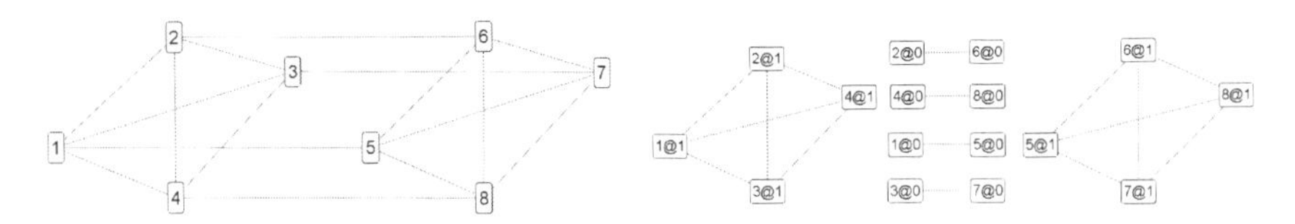

Fig. 4.17 Left: the 4-bipartite clique. Right: the disambiguated 4-bipartite clique

4.2.6 Hierarchical Divisive Chinese Whispers

In Section 4.1.1, the differences between a hierarchical clustering and a partition were laid out. CW in its presently discussed form is a partitioning algorithm that returns un-nested groups of vertices. To turn CW into a hierarchical cluster algorithm, one can split a cluster iteratively into smaller parts in a top-down way. Since a CW cluster is a set of vertices that receives a uniform class labelling and CW operates locally, a CW clustering of the subgraph induced by the cluster will most likely again result in a uniform labelling. To be able to still split this cluster, the subgraph has to be modified in a way that CW will be able to find more subclusters. A way of doing this for weighted graphs is to apply a threshold on edge weights and to delete edges that have a weight below that threshold. This can result in singleton vertices, which will not be contained in any of the subclusters, but remain assigned to the upper level cluster.

Especially for scale-free Small World graphs, hierarchical divisive CW provides a means to avoid one very large cluster. To illustrate this, the following experiment was carried out on the sentence-based significant word-cooccurrence graph of LCC's 1 million sentence German corpus (as used in Section 3.2.2): clustering this graph of 245,990 vertices with standard CW, the largest cluster consisted of 219,655 vertices. Only the largest cluster was selected for splitting: the edge weight threshold was set in a way that half of the vertices remained in the subgraph, the other half was left aside. In the first division step, clustering the subgraph of 109,827 vertices yielded a largest cluster of 44,814 vertices, of which again half of the vertices were selected for the second step. The largest cluster of this subgraph's partition (22,407 vertices) contained 16,312 vertices. In a similar third step, a subgraph of 8,157 vertices produced a largest cluster of 5,367 vertices.

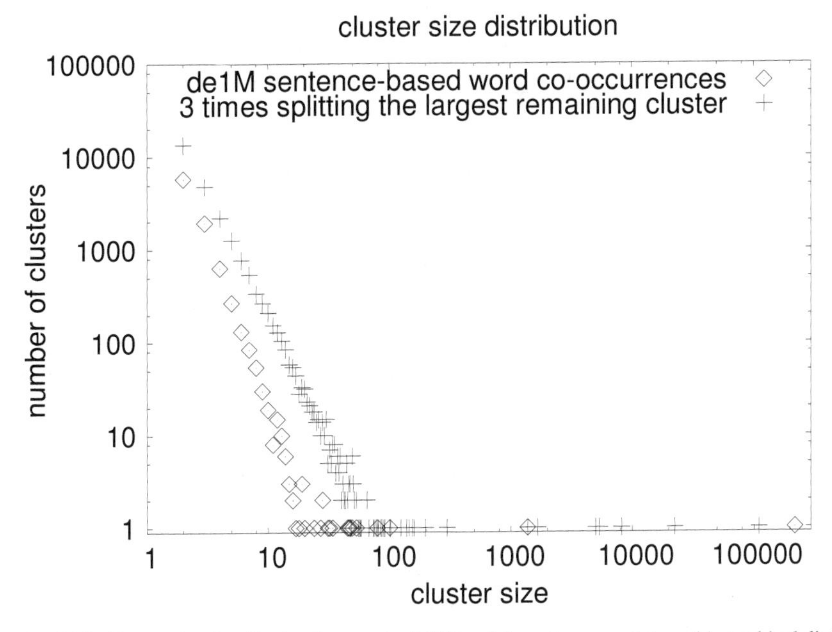

Fig. 4.18 Cluster size distribution for standard CW and three times applying a hierarchical divisive clustering step on the largest cluster

Figure 4.2.6 shows the cluster size distribution of the standard clustering and after the third divisive step. For the size distribution, the hierarchy of clusters is ignored. The quantitative analysis shows that the vertices of the very large initial cluster are separated in many small clusters after three times splitting the largest remaining cluster. The number of clusters increased from 9,071 in the initial clustering to 24,880 after the third step.

Deciding on which clusters to split and how many vertices to involve in the subcluster clustering render hierarchical divisive CW difficult to apply in a conclusive and motivated way. Easier to handle is the bottom-up variant of hierarchical clustering as presented in the next section.

4.2.7 Hierarchical Agglomerative Chinese Whispers

Section 4.2.4 described how the combination of several runs of CW runs results in a maximally fine-grained partition with respect to the iterative CW process. In applications, this partition might be too fine-grained, as having more clusters means less vertices per cluster, which might cause data sparseness problems in the application. As an example, consider lexical ambiguity: whereas CW, as other clustering algorithms, might find so-called micro-senses or facets [cf. 72], these fine-grained distinctions between e.g. *school* as a building or as an institution might not be useful in a word sense disambiguation task, or even harmful as there might not be enough clues for either micro-sense, yet the combination of their clues could suffice to distinguish them from unrelated usages.

Now a possibility is described how clusters can be arranged hierarchically in an agglomerative fashion. A partition of a graph can be viewed as dividing the graph's edges into inter-cluster and intra-cluster edges, the former connecting vertices in the same cluster whereas the latter being edges between clusters. Intuitively, two clusters have more in common, or are more similar to each other, if there a large amount of inter-cluster edges connects their vertices. In hierarchical agglomerative CW, hypervertices are formed by subsuming all vertices of one cluster under one hypervertex. The hyperedges connecting the hypervertices are weighted, their weight reflects the connection strength between the respective clusters. For example, a hyperedge weighting function can be given as

$$hyperweight(P_i, P_j) = \frac{|\{\{u,w\}|u \in P_i, w \in P_j\}|}{min(|P_i|, |P_j|)} \tag{4.11}$$

where P_i is the set of vertices constituting cluster i. This hyperweight function measures the amount of vertices in the smaller cluster that are endpoints of an edge between the two clusters. The resulting hypergraph — hypervertices and hyperedges — is again partitioned with CW, resulting in a set of hyperclusters. In this way, nested clusters are obtained by subordinating the clusters of the graph under the hyperclusters in the hypergraph. Iterative application results in a hierarchical clustering, with as many top-clusters as there are components in the graph. Figure 4.19 shows the hierarchical clustering of a graph in three steps.

As compared to other hierarchical clustering methods that mostly produce binary tree hierarchies, hierarchical agglomerative CW results in flat hierarchies with arbitrary branching factors. Apart from the instantiation of the *hyperweight* function, the hierarchical agglomerative version is parameter-free, and all modifications of CW can be employed for it.

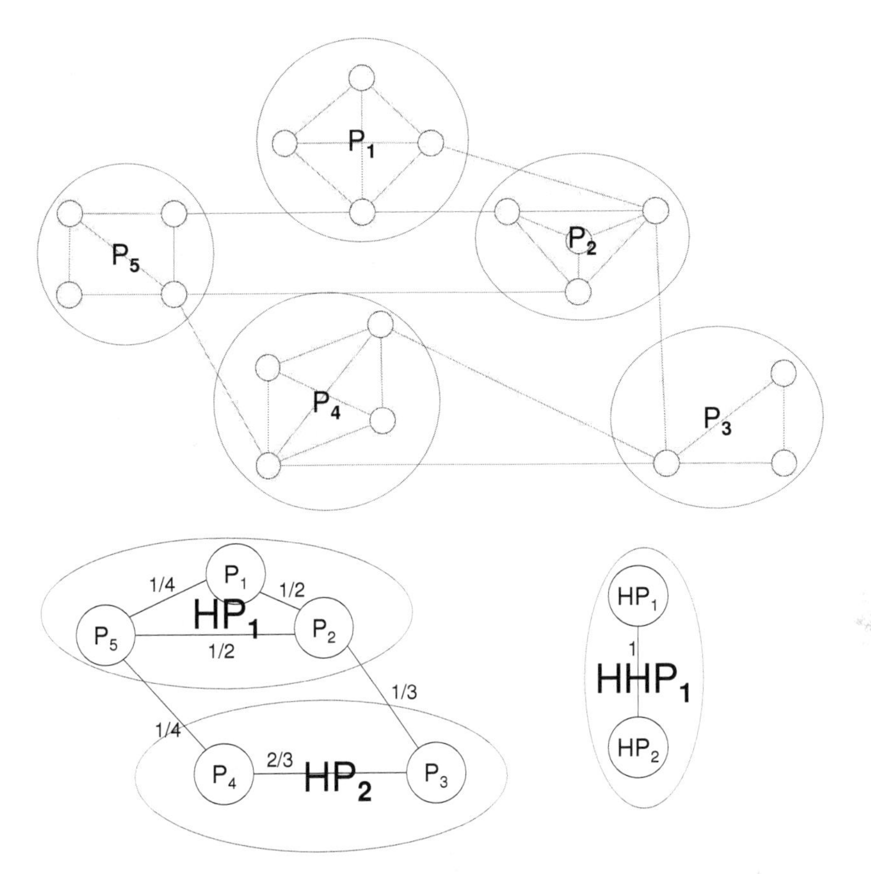

Fig. 4.19 Hierarchical agglomerative CW in three steps on an unweighted graph with 19 vertices. First step: five clusters P_1, P_2, P_3, P_4, P_5 by deterministic CW. Second step: two hyperclusters HP_1, HP_2. Third step: one large hyper-hypercluster HHP_1

4.2.8 Summary on Chinese Whispers

With its run-time linear in the number of edges, Chinese Whispers belongs to the class of graph partitioning algorithms at the lower bound of computational complexity: at least, the graph itself has to be taken into account when attempting to partition it, and the list of edges is the most compact form of its representation. The randomised nature of CW make it possible to run the algorithm in a decentralised and distributed way: since there are no constraints on the processing order of vertices and the algorithm uses only local information when updating the class of a vertex, a parallelisation is straightforward. This enables the partitioning of even larger graphs than the ones discussed here.

The property of standard CW to be parameter-free is highly desired for applications, as brittle parameter tuning can be avoided. Nevertheless, for datasets that demand a different treatment than standard CW provides, various extensions that try to leverage the non-deterministic nature and the absence of hierarchically nested clusters were laid out.

Since theoretical results are hard to obtain for randomised and non-deterministic algorithms, CW was tested on a variety of artificial graphs, showing great capabilities of producing meaningful clusters. Exploring the practical utility of CW will be undertaken in the next chapter.

An implementation of Chinese Whispers is available for download[5].

[5] http://wortschatz.informatik.uni-leipzig.de/~cbiemann/software/CW.html [June 1st, 2006].

Chapter 5
Unsupervised Language Separation

Abstract This chapter presents an unsupervised solution to language identification. The method sorts multilingual text corpora sentence-wise into different languages. In this attempt, the main difference to previous methods is that no training data for the different languages is provided and the number of languages does not have to be known beforehand. This application illustrates the benefits of a parameter-free graph clustering algorithm like Chinese Whispers, as the data — words and their statistical dependencies — are represented naturally in a graph, and the number of clusters (here: languages) as well as their size distribution is unknown. The feasibility and robustness of the approach for non-standard language data is demonstrated in a case study on Twitter data.

5.1 Related Work

According to Banko and Brill [15], Pantel et al. [190] and others, shallow methods of text processing can yield comparable results to deep methods when allowing them to operate on large corpora. The larger the corpus, however, the more difficult it is to ensure sufficient corpus quality. Most approaches in computational linguistics work on monolingual resources or on multilingual resources with monolingual parts, and will be disturbed or even fail if a considerable amount of 'dirt' (sublanguages or different languages) are contained. Viewing the World Wide Web as the world's largest text corpus, it is difficult to extract monolingual parts of it, even when restricting crawling to country domains or selected servers.

While some languages can be identified easily due to their unique encoding ranges in ASCII or UNICODE (like Greek, Thai, Korean, Japanese and Chinese), the main difficulty arises in the discrimination of languages that use the same encoding and some common words, as most of the European languages do. In the past, a variety of tools have been developed to classify text with respect to its language. The most popular free system, the TextCat Language Guesser[1] by Cavnar and Trenkle [54], makes use of the language-specific letter N-gram distribution and can determine 69 different natural languages. According to Dunning [85], letter trigrams can identify the language almost perfectly from a text length of 500 bytes on. Other language identification approaches use short words and high-frequency words as features, e.g. [132], or combine both approaches [cf. 216]. For a comparison, see [115].

All of these approaches work in a supervised way: given a sample of each language, the model parameters are estimated and texts are classified according to their similarity to the training texts. Supervised language identification is a solved problem, as reflected in the title of [163], where it is rendered as a suitable task for undergraduate exercises. But supervised training has a major drawback: the language identifier will fail on languages that are not contained in the training and, even worse, it will mostly have no clue about that and assign some arbitrary language. The reader is encouraged to apply supervised language identification systems to texts like "xx xxx x xxx ..." or "ATG GTT TCC CCT ACA ... " (genome sequences are usually an uncovered "language").

The method described here operates on words as features and finds the number of languages in a fully unsupervised way. Further, it is able to decide for every single sentence, to which language it belongs. Of course, it is not able to label text with the names of the involved languages, but rather groups sentences of the same language together. The method and some of the results have been published in [34].

5.2 Method

Employing word co-occurrence statistics (cf. Section 3.2.1), weighted graphs built from words as vertices and their associations as edges are constructed. Assuming monolinguality of sentences, there are more word pairs of the same language exhibiting significant co-occurrence than word pairs of different languages, Chinese Whispers will find one cluster for each language. The words in the clusters serve as features to identify the languages of the text collection by using a word-based language identifier.

As input, a multilingual, sentence-separated plain text corpus is assumed. The accuracy of the sentence separation, however, does not play an important role here and can be realised by splitting sentences on sentence delimiters like ".!?" or merely using paragraph boundaries. What matters is that the corpus is split into parts roughly corresponding to sentences in order to define the significant sentence-based co-occurrence graph as described in Section 3.2.1. Alternatively, window-based word co-occurrences could be employed, or the texts could be chopped in fixed-lenght sentence-like units of about 20 words[2].

[1] http://odur.let.rug.nl/~vannoord/TextCat/Demo/ [December 1st, 2006].

[2] the notion of words assumes tokenisation, which ideally would be another SD process.

A multilingual word co-occurrence graph is constructed by adding all word pairs that occur together at least twice with a significance of at least 1.64 (20% error; as preliminary experiments showed, however, the significance level can be set to any value in reasonable ranges without influencing results much), adding the words as vertices and weighting the edges by the significance value.

This multilingual word co-occurrence graph is clustered with standard Chinese Whispers, using 20 iterations. All clusters that exceed 1% of the size of the whole graph are used to define languages: the assumption is that words of the same language are found in the same cluster. Words that are contained in two or more languages will be assigned to only one cluster, languages are labelled by their cluster ID. The size threshold of 1% is arbitrary but showed to be a suitable setting to exclude noise in preliminary experiments.

These language clusters are used in the simple word-based language identifier described in [234] to assign cluster IDs to sentences: counting, how many words of the sentence are found in which cluster, the sentence is labelled with the ID of the cluster that contained most of the words. For labelling, at least two words and at least 10% of the words in the sentence have to be found in a cluster, otherwise the sentence is marked as 'unknown' — this was found to be a sensible setup to exclude questionable cases.

The algorithm for language separation that assigns a language ID $l(s)$ to every sentence s of corpus C is given in Algorithm 7:

Algorithm 7 langSepP(corpus C):

G = significant sentence-based word co-occurrence graph of C
partition $P = \mathrm{CW}(G)$
for all sentences $s = \{w_1, w_2, \ldots w_n\}$ in C **do**
  language $L = \arg\max_{P_i} |s \cap P_i|$
  if $(|L \cap s| < 2)$ OR $(|L \cap s| \times 10 < |s|)$ **then**
    $l(s)$=unknown
  else
    $l(s) = L$
  end if
end for

5.3 Evaluation

To test performance of the language separator, monolingual corpora of different languages are combined into a multilingual corpus. The success of the language separator is measured by the extent to which the original monolingual parts can be restored. For this, standard measures as used in information retrieval are employed:

- *Precision* (P) is the number of true positives divided by the sum of true positives and false positives,

- *Recall* (R) is the number of true positives divided by the total number of target items.

As it cannot be assumed beforehand that the number and extension of monolingual parts equals the number of language clusters, the following mapping between target languages and language clusters is carried out: a target language is mapped to the cluster ID that is found predominantly in the monolingual part of the target language. In case a cluster ID is found predominantly in several target languages, it is only mapped to the one target language where it constitutes the largest fraction, the other target languages get the next predominant cluster ID until a free cluster ID is found.

Table 5.1 shows an example of the mapping for a hypothetical dataset and exemplifies overall precision and recall as well as for single target languages: N = number of sentences in target language, the matrix $\{A,B,C,D\} \times \{1,2,3,4,5\}$ contains the number of sentences of language A..D labelled with cluster ID 1..5. Column 'map' indicates which cluster ID is mapped to which target language, columns P and R provide precision and recall for the single target languages. Here, language C gets assigned ID 4 from the clustering although the overlap between language C and cluster ID 2 is higher. The reason is that ID 2 is already mapped to language B, which has a higher percentual overlap to cluster ID 2 than language C. Notice that cluster ID 5 is not mapped to any language, which effects overall recall negatively.

Table 5.1 Exemplary mapping of target languages and cluster IDs. Overall precision: 0.6219, overall recall: 0.495

	N	1	2	3	4	5	map	P	R
A	100	75	0	0	0	0	1	1	0.75
B	200	5	110	40	1	20	2	0.625	0.55
C	300	5	120	60	30	10	4	0.133	0.1
D	400	15	0	280	25	0	3	0.875	0.7

5.4 Experiments with Equisized Parts for 10 Languages

In [34], experiments on mixing equisized parts to 7-lingual corpora are described. Here, the difficulty of the task is increased slightly in two ways: for experiments with equisized parts, 10 languages are used, and the sentences are not drawn subsequently from documents, but randomly from a large corpus. By randomly selecting sentences, the effect of word burstiness [cf. 64] is ruled out, which leads to fewer significant co-occurrences for small monolingual chunks: sentences from the same document tend to be centred around the document's topic which is reflected in their wording, producing significant co-occurrences of these topic-specific words. These should not be observed to that extent when selecting sentences randomly from documents encompassing a wide range of topics, making the task more difficult.

LCC was used as a data source to compile multilingual corpora consisting of Dutch, English, Finnish, French, German, Icelandic, Italian, Japanese, Norwegian and Sorbian up to 100,000 sentences per language.

Sorbian is a special case, as it is the only language that is not an official language of a country, but spoken by a Slavonic minority in Eastern Germany, why Sorbian texts contain words and even sentences in German. Considering this, it is not surprising that performance on Sorbian was lowest among all languages involved. Figure 5.1 shows overall results and results for Sorbian.

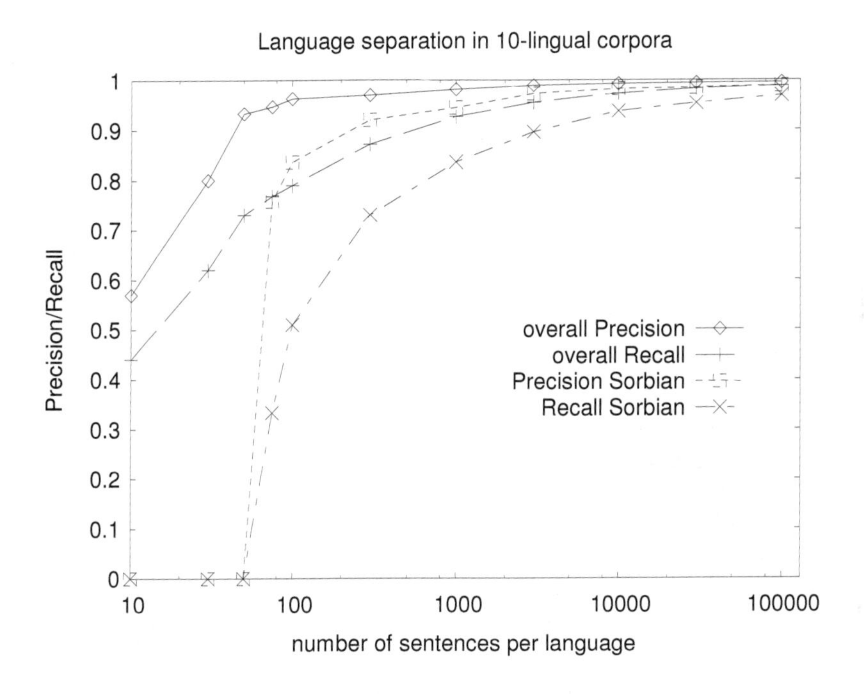

Fig. 5.1 Language separation in 10-lingual corpora with equisized monolingual parts for varying numbers of sentences

From 100 sentences per language on, the method is able to find exactly 10 clusters, and map them to the 10 languages involved. For fewer sentences, the number of clusters was smaller (7 for 10 sentences per language, 9 for 30 sentences per language). The easiest language to recognise in this experiment is Japanese, which shows a precision of 1 in all experiments and recall values from 0.9987 to 1, which can be attributed to its distinct character set. The second easiest language is English, scoring more than 0.9997 on precision and more than 0.997 on recall for all experiments with 1,000 sentences per language or more. In the Sorbian case, a manual inspection revealed that about 1% of sentences in this data are either proper German or German dialect, sometimes mixed with Sorbian.

Most misclassifications are caused by embedded foreign-language elements, such as the sentence *"INA besitzt Anteile an der Banca Nazionale di Lavoro und am Banco di Napoli."* in the German part that is misclassified as Italian because it contains names of two Italian banks. Languages of the same language family are more difficult to separate, e.g. Norwegian-Icelandic or Dutch-German. Most confusion is observable for English, which is the largest contributor of foreign-language material in all other languages. Table 5.2 shows the confusion matrix for the experiment with 100,000 sentences per language.

cluster ID	1	2	3	4	5	6	7	8	9	10
French	99709	13			3	3			1	
English	19	99950			3	1	2			
Icelandic	26	118	99060		17	13	30	17	2	
Japanese				99993						
Italian	29	34	1	1	99451	2			1	3
Norwegian	56	158	668	4	8	97165	27	4	36	4
German	21	117	1	1	6		99235		7	2
Finnish	14	21	5	1	4	60	3	99616	2	2
Dutch	611	999	2	1	30	6	273	4	96793	5
Sorbian	36	138	3	2	13	7	920	14	4	96909

Table 5.2 Confusion matrix of clusters and languages for the experiment with 100,000 sentences per language

Recall is dependent on sentence length: in short sentences, it is less likely to find enough words from the clusters. This is depicted in Figure 5.2: sentences longer than 10 words almost surely get classified.

Most long sentences marked as 'unknown' contained names, dates or address parts, such as e.g.

- *Ruth Klink, Niddatal 1, 100,- ; Margot Klöck, 50,- ; Frank Klöß, Bad Vilbel, 20,-.* (German)
- *BC-Net Secretariat , European Commission , DG23 , Rue de la Loi 200 , B-1049 Brussels .* (English)
- *Aspelinin veli) 22 - Aspelin Emil (JRA:n serkku) 10 - Aspelin Henrik Emanuel (Manne , JRA:n veli) 1 Hr19 J.R.* (Finnish)

These kind of sentences are not informative regarding language structure and it might be a good idea to remove them during corpus compilation anyway.

Unsupervised language separation proves capable of separating equisized monolingual chunks from each other. Performance is almost perfect starting from a few hundred sentences per language. The next section is devoted to testing the stability of the method when applying it to bilingual mixtures with monolingual parts of different sizes.

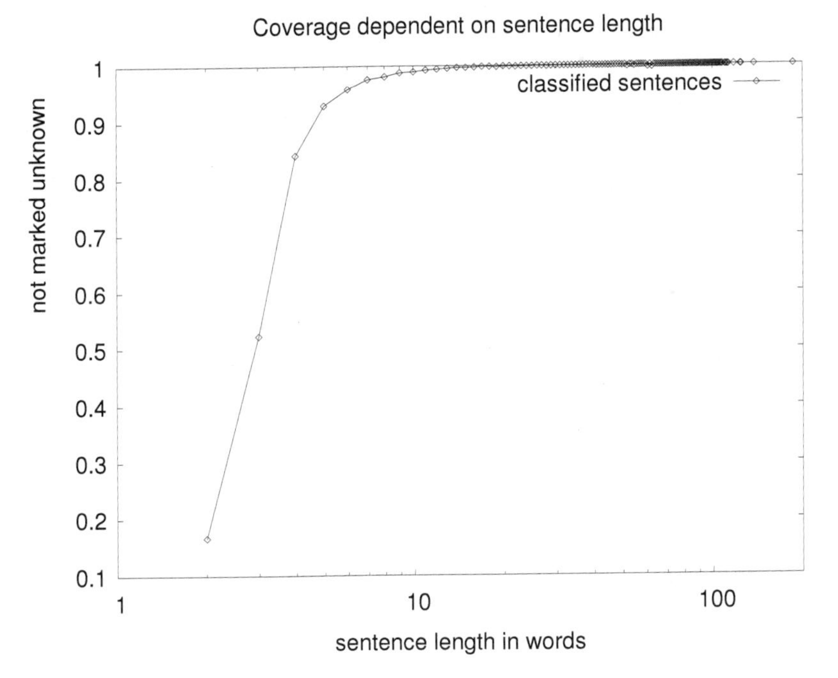

Fig. 5.2 Coverage dependent on sentence length

5.5 Experiments with Bilingual Corpora

While the experiments in the previous section focussed on equal fractions of many languages, the method is now evaluated on bilingual corpora with monolingual fractions of differing sizes. This setting is somewhat more realistic when identifying languages in web data, for a top-level domain usually provides most of its content in one language, and the task is to remove substantially smaller parts of foreign language material. The experimental setup is as follows: into a monolingual corpus of 100,000 sentences, chunks of 10, 30, 100, 300, 1,000, 3,000, 10,000 and 30,000 sentences of another language are injected. The major language is always identified; therefore performance is only reported on the injected language.

In the previous section it was observed that languages of the same language family are harder to separate than very different languages or even languages using different alphabets. Therefore, two extreme cases are examined here: mixtures of French as major and Japanese as minor language, and mixtures of Swedish as major and Norwegian as a minor language. Separation of other language pairs can be expected to perform in between these borderline cases.

French and Japanese do not have any words in common apart from names; therefore the multilingual word graph is easily separable. The only crucial parameter is the cluster size threshold: setting it to 1% as in the previous section, the

Japanese part is recognised perfectly from 300 sentences on. Lowering the threshold to 0.03%, it is even possible to find every single of the 10 sentences of Japanese in a 100,000 sentence corpus of French without a drop on precision and recall in the French part.

In [34], experiments on injecting English in an Estonian corpus and Dutch in a German corpus succeeded in almost perfect separation from 500 minor language sentences on. It was only slightly more difficult to find the French injection into an Italian corpus due to the relatedness of these languages: the more words languages have in common, the more difficult it is to separate them. In [197], the overlap between the most frequent 1,000 words of several European languages is given. Danish and Norwegian are the most similar languages with respect to this measure, as they share almost half of their top 1,000 words. Experiments aiming at separating Danish and Norwegian mixtures with the method described here did not succeed. The next most similar language pair is Swedish and Norwegian with sharing over 1/4th of these words. As the results in Figure 5.3 show, this results in lower performance on this pair as compared to the multilingual experiments.

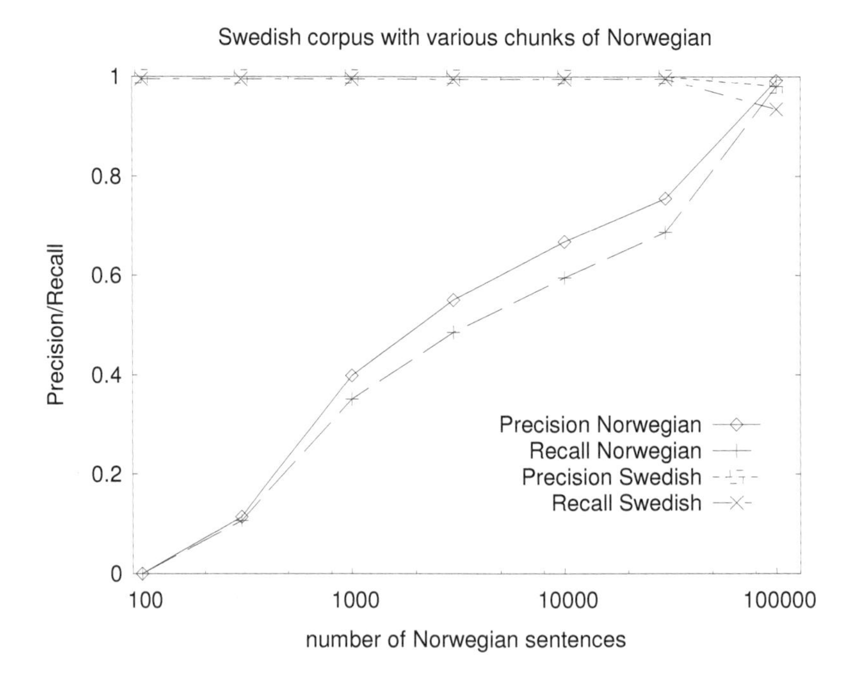

Fig. 5.3 Language separation performance for Norwegian injected in a Swedish corpus of 100,000 sentences

While performance on 100,000 sentences of each Norwegian and Swedish is comparable to what has been reported in the equisized experiments, smaller fractions of Norwegian are difficult to identify and get often labelled with the same ID as the larger Swedish part.

5.6 Case study: Language Separation for Twitter

Structure Discovery processes are especially suited in situations, where the data does not follow standard conventions. One example where non-standard language is used extensively is Twitter[3], a micro-blogging service where publicly available broadcast messages (called *tweets*) are limited to a mere 140 characters. This limitation causes users to be very creative in shortening words, using abbreviations and emoticons. Furthermore, special word classes exist that mark users (starting with @) or self-defined tags (starting with #), and many tweets contain a URL, which is usually shortened.

Here are some examples of tweets from March 26, 2010 that contain non-standard English.

- *RT @Pete4L: Guys plz d/l the lettr Ive written 2 Jeff Gaspin, he's THE man who can giv us #Heroes S5 http://tinyurl.com/y9pcaj7 #Heroes100*
- *@SkyhighCEO LOOOL heyyy! shullup! #Jujufish*
- *LUV HER o03.o025.o010 thanx to da sis ariana 4 makin da pic I most def lyk it but goin 2 da rink 2 moRrow ya dawg wit da http://lnk.ms/5svJB*
- *Q: hay justin SCREEEEEEEEEM!!!!!!! i luv u OMG!!!!!!!!! i did a quiz ubout if me and u wer thu only ones o http://www.society.me/q/29910/view*

This type of language is not a fringe phenomenon, but can be encountered frequently in Twitter streams. Whereas most of the long examples contain enough stopwords to classify them as English, the second example does not contain any valid English word. In preliminary explorations it was observed that the language and geo tag provided with the tweet only correlates weakly with its language.

Other language processing systems for Twitter usually assume their tweets at least to be monolingual. Attempts have been made to normalise the spelling [122] and to set up a specialised part-of-speech-tagger [109]. Here, a pre-processing step is described that increases the recall of tweets for these methods.

The language separation method was executed on a corpus of one full day (March 10, 2010) of tweets, comprising over 38 million tweets. 35 clusters were identified in this data. The largest cluster represents English, consisting of 1.1 million words. All words of the second tweet above are contained in this cluster. Most of the clusters actually correspond to different languages; there is also a cluster of location names which co-occur frequently in Dutch traffic updates, which are recognised as a sublanguage. For some large twitter languages like English, Portuguese and Spanish, there are separate clusters for all-uppercase tweets. Indonesian and Malay, languages even more similar to each other than Danish and Norwegian, could not be

[3] http://www.twitter.com [August 31, 2011]

distinguished. Further, the system as applied out-of-the-box was not able to discriminate between Chinese and Japanese, which could be fixed easily based on Unicode code pages. Despite the shortness of messages and the non-standard language use, only about 4% of tweets were not assigned to any language cluster.

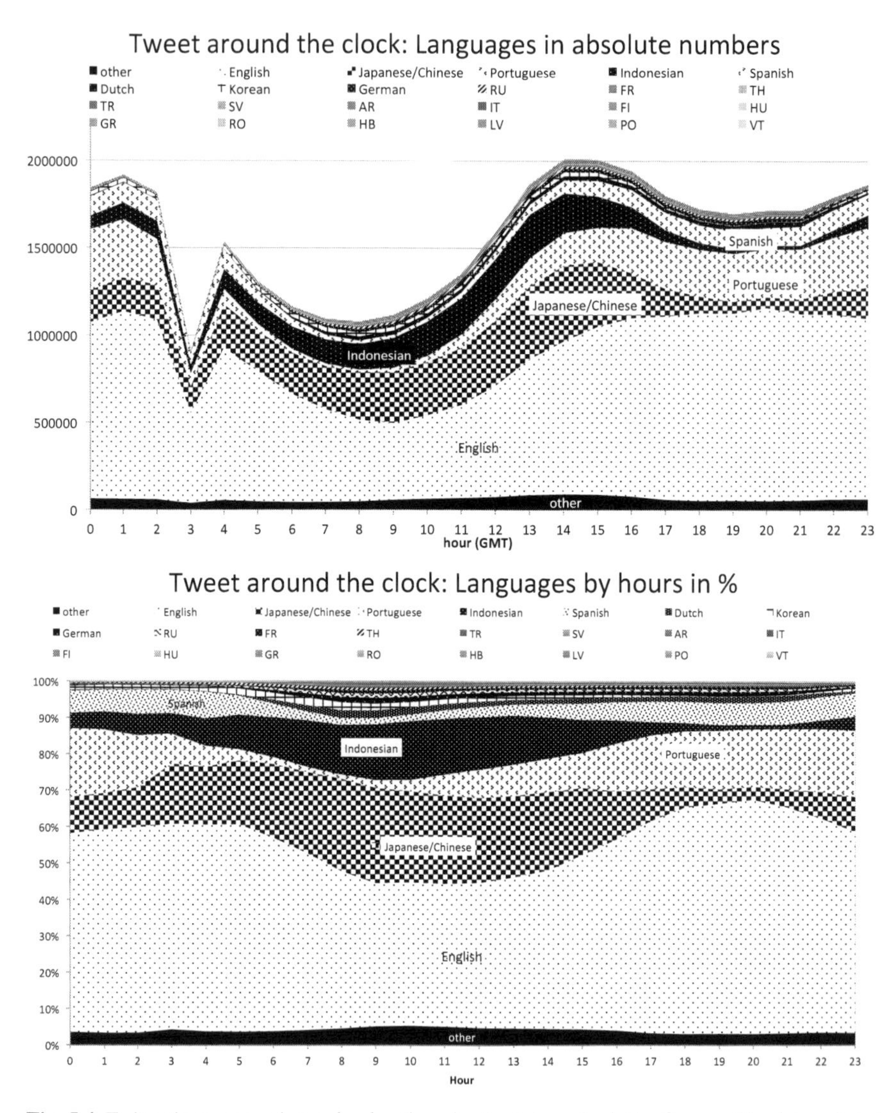

Fig. 5.4 Twitter language volumes by day time, in absolute and relative figures. The data point at 2:00-3:00 GMT is distorted because of an outage.

In Figure 5.4, the volume of the 23 largest languages is plotted per hour, according to the tweet's time stamp. The absolute numbers correlate to the population density of the world per time zone. This data analysis reveals that Twitter is very popular in Brazil and in South East Asia (Indonesia/Malaysia), and the peaks in these respective regions happen at different time zones. Even with a skewed distribution of languages, the system reliably recognises smaller languages such as Vietnamese and Latvian. The unsupervised language separation approach has been demonstrated to work well in practice.

5.7 Summary on Language Separation

Apart from the experiments with Scandinavian languages, which aimed at testing the method for very extreme cases, it is possible to conclude that unsupervised language separation arrives at performance levels that are comparable to its supervised counterparts. The problem of language separation and identification can be regarded as solved not only from a supervised, but also from a Structure Discovery perspective.

To collect the few unclassified sentences, the approach could be extended to the document level: unknown sentences are assigned the language that surrounds them. Using the extensions of CW as discussed in Sections 4.2.4 and 4.2.5 does not seem necessary in the light of the obtained performance levels.

An implementation of this method called "langSepP" (language separation program) is available for download[4].

[4] http://wortschatz.informatik.uni-leipzig.de/~cbiemann/software/langSepP.html [July 7th, 2007]

Chapter 6
Unsupervised Part-of-Speech Tagging

Abstract In this chapter, homogeneity with respect to syntactic word classes (parts-of-speech, POS) is aimed at. The method presented in this section is called unsupervised POS-tagging, as its application results in corpus annotation in a comparable way to what POS-taggers provide. Nevertheless, its application results in slightly different categories as opposed to what is assumed by a linguistically motivated POS-tagger, which hampers evaluation methods that compare unsupervised POS tags to linguistic annotations. To measure the extent to which unsupervised POS tagging can contribute in application-based settings, the system is evaluated in supervised POS tagging, word sense disambiguation, named entity recognition and chunking, improving on the state-of-the-art for supervised POS tagging and word sense disambiguation. Unsupervised POS-tagging has been explored since the beginning of the 1990s. Unlike in previous approaches, the kind and number of different tags is here generated by the method itself. Another difference to other methods is that not all words above a certain frequency rank get assigned a tag, but the method is allowed to exclude words from the clustering, if their distribution does not match closely enough with other words. The lexicon size is considerably larger than in previous approaches, which results in a more robust tagging.

6.1 Introduction to Unsupervised POS Tagging

Assigning syntactic categories to words is an important pre-processing step for most NLP applications. POS tags are used for parsing, chunking, anaphora resolution, named entity recognition and information extraction, just to name a few.

Essentially, two things are needed to construct a tagger: a lexicon that contains tags for words and a mechanism to assign tags to tokens in a text. For some words, the tags depend on their use, e.g. in "I saw the man with a saw". It is also necessary to handle previously unseen words. Lexical resources have to offer the possible tags, and a mechanism has to choose the appropriate tag based on the context, in order to produce annotation like this: "I/PNP saw/VVD the/AT0 man/NN1 with/PRP a/AT0 saw/NN1 ./PUN"[1].

Given a sufficient amount of manually tagged text, two approaches have demonstrated the ability to learn the instance of a tagging mechanism from labelled data and apply it successfully to unseen data. The first is the rule-based approach [47], where transformation rules are constructed that operate on a tag sequence delivered by the lexicon. The second approach is statistical, for example HMM-taggers [57, inter al.] or taggers based on conditional random fields [see 148]. Both approaches employ supervised learning and therefore need manually tagged training data. Those high-quality resources are typically unavailable for many languages and their creation is labour-intensive. Even for languages with rich resources like English, tagger performance breaks down on noisy input. Texts of a different genre than the training material may also create problems, e.g. emails as opposed to newswire or literature. It is, in general, not viable to annotate texts for all these cases.

Here, an alternative needing much less human intervention is described. Steps are undertaken to derive a lexicon of syntactic categories from unstructured text following the Structure Discovery paradigm. Hence, it is not possible to aim at exact correspondence with linguistically motivated tagsets, but for obvious reasons: even for the same language, linguistically motivated tagsets differ considerably, as it was measured for various tagsets for English by Clark [66].

Two different techniques are employed here, one for high-and medium frequency words, another for medium- and low frequency words. The categories will be used for the tagging of the same text the categories were derived from. In this way, domain- or language-specific categories are automatically discovered. Extracting syntactic categories for text processing from the texts to be processed fits the obtained structures neatly and directly to them, which is not possible using general-purpose resources.

With moving POS tagging to a data-driven, unsupervised step that can serve as feature-based input for subsequent steps, a major step in alleviating the acquisition bottleneck can be taken. The motivation behind this work is primarily to lower the amount of work that goes into manual annotation or the creation of rule sets; on a larger perspective, however, it can also unveil principles of language structure in such as common features and differences between languages are mirrored in the way the data arranges itself for different languages.

The contents of this chapter have also been published in [29].

6.2 Related Work

There are a number of approaches to derive syntactic categories. All of them employ a syntactic version of Harris' distributional hypothesis [123]: words of similar parts of speech can be observed in the same syntactic contexts. Measuring to what extent

[1] in this tagset [106], PNP stands for personal pronoun, VVD is full verb, AT0 is determiner is singular or plural, NN1 is singular noun, PRP is Preposition, PUN is punctuation.

two words appear in similar contexts measures their similarity [cf. 179]. As function words form the syntactic skeleton of a language and almost exclusively contribute to the most frequent words in a corpus, contexts in that sense are often restricted to the most frequent words. The words used to describe syntactic contexts are further called *feature words*. *Target words*, as opposed to this, are the words that are to be grouped into syntactic clusters. Note that usually, the feature words form a subset of the target words.

The general methodology [98; 218; 220; 107; 65; 203] for inducing word class information can be outlined as follows:

1. Collect global context vectors of target words by counting how often feature words appear in neighbouring positions.
2. Apply a clustering algorithm on these vectors to obtain word classes

Throughout, feature words are the 150-250 words with the highest frequency. Some authors employ a much larger number of features and reduce the dimensions of the resulting matrix using Singular Value Decomposition [218; 203]. The choice of high frequency words as features is motivated by Zipf's law: these few stop words constitute the bulk of tokens in a corpus. Pruning context features to these allows efficient implementations without considerably losing on coverage. Contexts are the feature words appearing in the immediate neighbourhood of a word. The word's global context is the sum of all its contexts. Figure 6.1 illustrates the collection of contexts for a German toy example.

```
... COMMA sagte der Sprecher bei der Sitzung FULLSTOP
... COMMA rief der Vorsitzende in der Sitzung FULLSTOP
... COMMA warf in die Tasche aus der Ecke FULLSTOP
```

Features: der(1), die(2), bei(3), in(4), FULLSTOP (5), COMMA (6)

position	-2				-1						+1						+2	
target/feature	3	4	5	6	1	2	3	4	5	6	1	2	3	4	5	6	1	2
sagte										1	1							
rief										1	1							
warf										1				1				1
Sprecher					1								1				1	
Vorsitzende					1									1			1	
Tasche		1				1											1	
Sitzung	1	1			2										2			
Ecke					1										1			

Fig. 6.1 Corpus and context vectors for 6 feature words and a context window of size 4. The feature vectors of different positions are concatenated

As outlined in Chapter 4, clustering consists of a similarity measure and a clustering algorithm. Finch and Chater [98] use the Spearman Rank Correlation Coefficient and a hierarchical clustering, Schütze [218, 220] uses the cosine between vector angles and Buckshot clustering, Gauch and Futrelle [107] use cosine on Mutual Information vectors for hierarchical agglomerative clustering and Clark [65] applies Kullback-Leibler divergence in his CDC algorithm.

An extension to this generic scheme is presented in [66], where morphological information is used for determining the word class of rare words. Freitag [104] does not sum up the contexts of each word in a context vector, but uses the most frequent instances of four-word windows in a co-clustering algorithm [81]: rows and columns (here words and contexts) are clustered simultaneously. Two-step clustering is undertaken by Schütze [218]): clusters from the first step are used as features in the second step.

The number of target words in the clustering differ from 1,000 target words in a 200,000 token corpus [107] over 5,000 target words [98; 104] to all 47,025 words in the Brown Corpus in [220]. Clark [65] uses 12 million tokens as input; Finch and Chater [98] operate on 40 million tokens.

Evaluation methodologies differ considerably amongst the papers discussed here. Finch and Chater [98] inspect their clusters manually, Rapp [203] performs flawlessly in sorting 50 medium frequency words into nouns, verbs and adjectives. Schütze [220] presents precision and recall figures for a reduced tagset, excluding rare and non-English-word tags from the evaluation. More recent approaches [65; 66; 104] employ information-theoretic measures, see Section 6.8. Regarding syntactic ambiguity, most approaches do not deal with this issue while clustering, but try to resolve ambiguities at the later tagging stage.

As the virtue of unsupervised POS tagging lies in its possible application to all natural languages or domain-specific subsets, it is surprising that in most previous works, only experiments with English are reported. An exception is [66], who additionally uses languages of the Slavonic, Finno-Ugric and Romance families.

A problem with most clustering algorithms is that they are parameterised by the number of clusters. As there are as many different word class schemes as tagsets, and the exact amount of word classes is not agreed upon intra- and interlingually, having to specify the number of desired clusters a-priori is clearly a drawback. In that way, the clustering algorithm is forced to split coherent clusters or to join incompatible sub-clusters. In contrast, unsupervised part-of-speech induction means the induction of the tagset, which implies finding the number of classes in an unguided way.

Another alternative which operates on a predefined tagset is presented by Haghighi and Klein [119]: in this semi-supervised framework, only three words per tag have to be provided to induce a POS-tagger for English with 80% accuracy. The amount of data the authors use in their experiments is rather small (8,000 sentences), but their computationally expensive methods — gradient-based search to optimise Markov Random Field parameters — does not allow for substantially more input data. For methods along the same line of research — computationally expensive methods that aim to get the most out of comparatively small amounts of data for POS induction — the reader is referred to [114] and [237].

6.3 System Architecture

Input to the system is a considerable amount of unlabelled, tokenised monolingual text without any POS information. In a first stage, Chinese Whispers is applied to distributional similarity data, which groups a subset of the most frequent 10,000 words of a corpus into several hundred clusters (tagset 1). Second, similarity scores on neighbouring co-occurrence profiles are used to obtain again several hundred clusters of medium- and low frequency words (tagset 2). The combination of both partitions yields sets of word forms belonging to the same induced syntactic category. To gain on text coverage, ambiguous high-frequency words that were discarded for tagset 1 are added to the lexicon. Finally, a Viterbi trigram tagger is trained with this lexicon and augmented with an affix classifier for unknown words.

Figure 6.2 depicts the architecture of the process of unsupervised POS-tagging from unlabelled to fully labelled text. The following sections will outline the components in more detail.

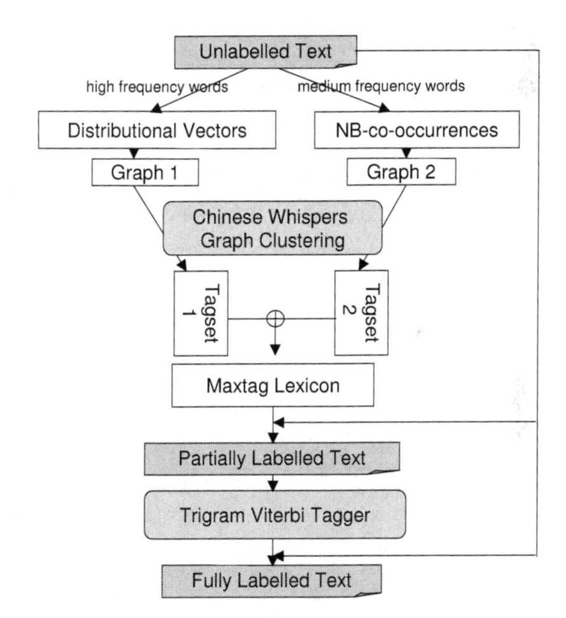

Fig. 6.2 Diagram of the process of unsupervised POS tagging, from unlabelled over partially labelled to fully labelled text

The method employed here follows the coarse methodology as described in the previous subsection, but differs from other works in several respects. Although four-word context windows and the top frequency words as features are used [as in 220], the cosine similarity values between the vectors of target words are transformed into a graph representation in order to be able to cluster them with CW. Additionally, a method to identify and incorporate POS-ambiguous words as well as low-frequency words into the lexicon is implemented.

6.4 Tagset 1: High and Medium Frequency Words

Four steps are executed in order to obtain tagset 1 for high- and medium frequency words from a text corpus.

1. Determine 10,000 target and 200 feature words from frequency counts
2. Collect context statistics and construct graph
3. Apply Chinese Whispers on graph.
4. Add the feature words not present in the partition as one-member clusters.

The graph construction in step 2 is conducted by adding an edge between two words a and b with weight[2] $w = 1/(1 - cos(\overrightarrow{a}, \overrightarrow{b}))$, computed using the feature vectors $\overrightarrow{a}$ and $\overrightarrow{b}$ (cf. Figure 6.1) of words a and b. The edge is only drawn if w exceeds a similarity threshold s. The latter influences the number of words that actually end up in the graph and get clustered. It might be desired to cluster fewer words with higher confidence as opposed to running the risk of joining two unrelated clusters because of too many ambiguous words that connect them. After step 3, there is already a partition of a subset of target words that can be perceived as tagset. Figure 6.3 shows the weighted graph and its CW-partition for the example given in Figure 6.1. The number of target words is limited by computational considerations: since the feature vectors have to be compared in a pair-wise fashion, a considerably higher number of target words results in long run times. The number of feature words was examined in preliminary experiments, showing only minor differences with respect to cluster quality in the range of 100–300.

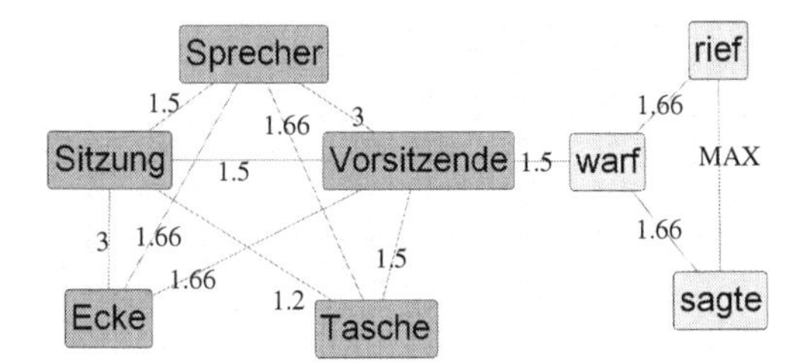

Fig. 6.3 Graph for the data given in Figure 6.1 and its partition into nouns and verbs

As noted e.g. in [220], the clusters are motivated syntactically as well as semantically and several clusters for each open word class can be observed. The distinctions are normally finer-grained than existing tagsets, as Figure 6.4 illustrates.

Since the feature words form the bulk of tokens in the corpus, it is clearly desired to make sure that they appear in the tagset, although they might end up in clusters with only one element. This might even be desired, e.g. for English *'not'*, which usually has its own POS-tag in linguistic tagsets. This is done in step 4, where assigning separate word classes for high frequency words is considered to be a more

---

[2] cosine similarity is a standard measure for POS induction, however, other measures would be possible.

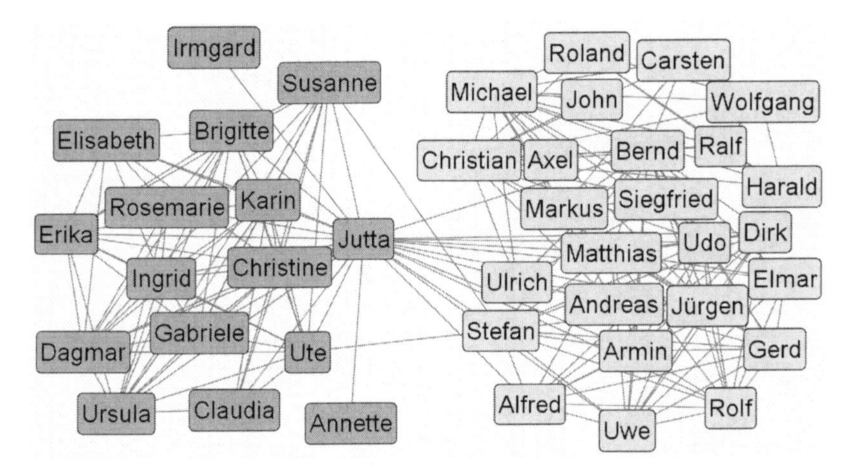

Fig. 6.4 Fine-grained distinctions: female and male first names from German corpus. The figure shows only a local neighbourhood of the graph for tagset 1

robust choice than trying to disambiguate them while tagging. Starting from this, it is possible to map all words contained in a cluster onto one feature and iterate this process, replacing feature words by the clusters obtained [cf. 218]. In that way, higher counts in the feature vectors are obtained, which could provide a better basis for similarity statistics. Preliminary experiments showed, however, that only marginal improvements could be reached, as text coverage is not substantially increased.

Table 6.1 shows a selection of clusters for the BNC, including several different clusters for nouns. Evaluating lexical clusters against a gold standard may lead to inconclusive results, because the granularities of the gold standard and the clusters usually differ, e.g. English singular and plural nouns end up in one cluster, but first and last names are distinguished. The evaluation scores are largely depending on the tagset used for gold standard. Here, an information-theoretic measure is employed that allows an intuitive interpretation: Entropy precision (EP) measures the extent to which the gold standard classification is reconstructable from the clustering result. EP directly relates to the precision measure in information retrieval. Its counterpart, recall as the number of retrieved vs. the total number of instances relates to the coverage on target words as reached by the clustering algorithm. For the gold standard, each word gets assigned its most frequent tag, ignoring POS-ambiguities. Despite all these disadvantages, EP provides a means to relatively compare the quality of partitions for varying thresholds s.

Definition Entropy Precision (EP): Let $G = G_1, ...G_m$ be the gold standard classification and $C = C_1, ...C_p$ be the clustering result. Then, EP is computed as follows:

$$EP(C,G) = \frac{M_{CG}}{I_G} \tag{6.1}$$

with mutual information M_{XY} between X and Y

$$M_{XY} = \sum_{xy} P(x,y) ln \frac{P(x,y)}{P(x)P(y)} \tag{6.2}$$

and I_X entropy of X.

$$I_X = -\sum_{x} P(x) ln P(x) \tag{6.3}$$

A maximal EP of 1 is reached by a trivial clustering of singleton clusters. This does not impose a severe problem, considering the typical cluster size distribution as depicted in Figure 6.5. Nevertheless, optimising EP promotes a large number of small clusters, which is why the number of clusters has to be provided along with the EP figures to give an impression of the result's quality. A minimal EP of 0 indicates statistical independence of C and G.

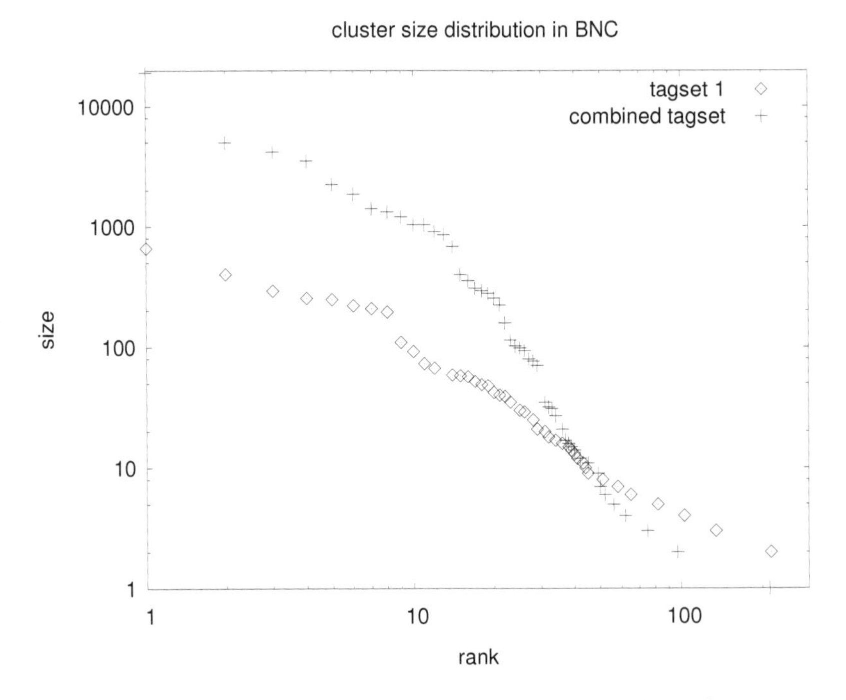

Fig. 6.5 Cluster size distribution for tagset 1 and combined tagset in the BNC

For evaluation of tagset 1, three corpora of different languages were chosen: 10 million sentences of German tagged with 52 tags using TreeTagger [215], the 6 million sentences of BNC for English, pretagged semi-automatically with the CLAWS tagset of 84 tags [106] and 1 million sentences from a Norwegian web corpus tagged with the Oslo-Bergen tagger [118], using a simplified tagset of 66 tags. Figure 6.6 gives the EP results for varying numbers of target words included in the partition and the number of clusters.

rank	size	gold standard tags (count)	description	sample words
1	662	NN1(588), NN2(44)	Singular Nouns	day, government, world, system, company, house, family
2	401	NN1(311), NN2(86)	Singular Nouns	part, end, state, development, members, question, policy, ...
3	292	NN2(284), NN1(7)	Plural Nouns	men, services, groups, companies, systems, schools, ...
4	254	NP0(252), NN1(2)	First Names	John, David, Peter, Paul, George, James, Michael, ...
5	247	AJ0(233), NN1(9)	Adjectives	social, political, real, economic, national, human, private, ...
6	220	NN1(148), NN2(63)	Singular and Plural Nouns	business, water, service, staff, land, training, management, ...
7	209	VVI(209)	Verbs	get, make, take, give, keep, provide, play, move, leave, ...
8	195	AJ0(118), NN1(25)	Adjectives (country)	British, police, New, European, individual, National, ...
9	110	NP0(109), NN1(1)	Last names	Smith, Jones, Brown, Wilson, Lewis, Taylor, Williams, ...
10	92	AJ0(89), CRD(1)	Adjectives (size/quality)	new, good, little, few, small, great, large, major, big, special
11	73	AJ0(73)	Adjectives (animate)	heavy, beautiful, quiet, soft, bright, charming, cruel, ...
12	67	NN2(67)	Plural Nouns	problems, conditions, costs, issues, activities, lines, ...
12	67	NP0(66), NN1(1)	Countries	England, Scotland, France, Germany, America, Ireland, ...
16	57	NP0(57)	Cities	Oxford, Edinburgh, Liverpool, Manchester, Leeds, Glasgow, ...
22	39	AV0(39)	Sentence Beginning	Well, However, Thus, Indeed, Also, Finally, Nevertheless, ...
25	30	NN2(30)	Plural Professions	teachers, managers, farmers, governments, employers, ...
34	17	CRD(17)	Numbers	three, four, five, six, ten, eight, seven, nine, twelve, fifteen, ...
65	6	NP0(6)	Titles	Mr, Mrs, Dr, Miss, Aunt, Ms
217	2	AT0(2)	Indefinite determiner	a, an
217	2	NP0(2)	location 1st	Saudi, Sri
217	2	VVZ, VVD	to wear	wore, wears
217	2	VVZ, VVD	to insist	insisted, insists

Table 6.1 Selected clusters from the BNC clustering for setting s such that the partition contains 5,000 words. In total, 464 clusters are obtained. EP for this partition is 0.8276. Gold standard tags have been gathered from the BNC

From Figure 6.6 it is possible to observe that EP remains stable for a wide range of target word coverage between about 2,000-9,000 words. The number of parts is maximal for the medium range of coverage: at higher coverage, POS-ambiguous words that are related to several clusters serve as bridges. If too many links are established between two clusters, CW will collapse both into one cluster, possibly at cost of EP. At lower coverage, many classes are left out. This evaluation indicates the language-independence of the method, as results are qualitatively similar for all languages tested.

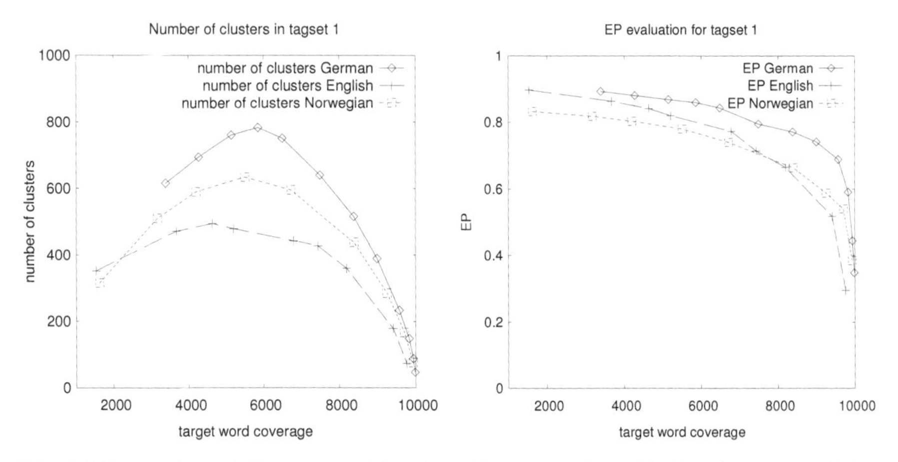

Fig. 6.6 Tagset size and Entropy precision dependent on number of included target words for tagset 1

As indicated above, lexicon size for tagset 1 is limited by the computational complexity of step 2, which is time-quadratic in the number of target words. Due to the non-sparseness of context vectors of high-frequency words there is not much room for optimisation. In order to add words with lower frequencies, another strategy is pursued.

6.5 Tagset 2: Medium and Low Frequency Words

As noted in [86], log likelihood statistics capture word bigram regularities. Given a word, its neighbouring co-occurrences as ranked by their log likelihood ratio are the typical immediate contexts of the word. Regarding the highest ranked neighbours as the profile of the word, it is possible to assign similarity scores between two words A and B according to how many neighbours they share, i.e. to what extent the profiles of A and B overlap. The hypothesis here is that words sharing many neighbours should usually belong to the same part-of-speech. For the acquisition of word classes in tagset 2, the second-order neighbouring co-occurrence graph is used, cf Section

3.2.2. To set up the graph, a co-occurrence calculation is performed which yields word pairs based on their significant co-occurrence as immediate neighbours. Here, all word pairs exceeding a log likelihood threshold of 2.71 (corresponding to 10% error, yet the outcome is robust in a wide threshold range) enter this bipartite graph. Note that if similar words occur in both parts, they form two distinct vertices. Only words with a frequency rank higher than 2,000 are taken into account: in preliminary experiments, high-frequency words of closed word classes spread over the clusters, resulting in deteriorated tagging performance later, so these are excluded.

This graph is transformed into a second-order graph by comparing the number of common right and left neighbours for two words. The similarity (edge weight) between two words is the sum the number of common neighbours on both sides. Figure 6.7 depicts the significant neighbouring graph, the second-order graph derived from it, and its CW-partition. The word-class-ambiguous word 'drink' (to drink the drink) is responsible for all inter-cluster edges. In the example provided in Figure 6.7, three clusters are obtained that correspond to different parts-of-speech. For computing the similarities based on the significant neighbour-based word co-occurrence graphs for both directions, at maximum the 200 most significant co-occurrences per word are considered, which regulates the density of ther graph and leads to improvements in run-time.

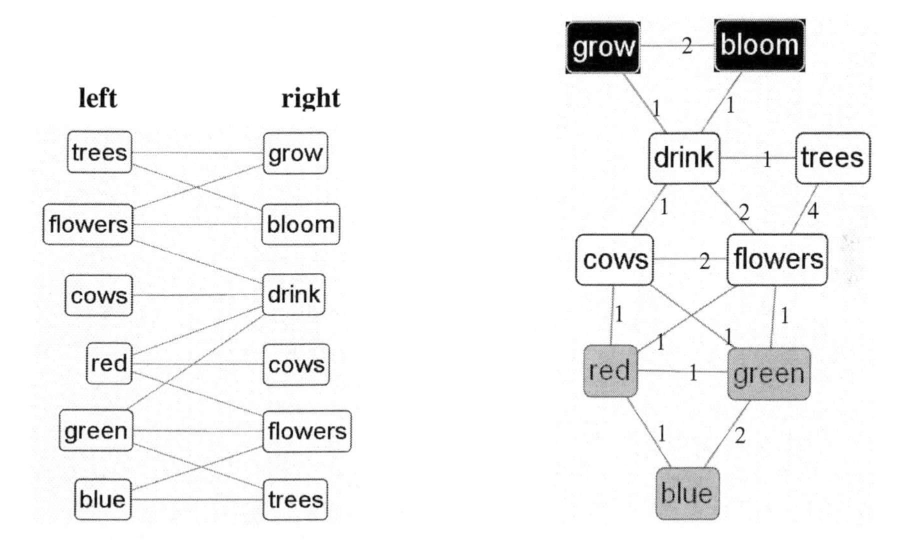

Fig. 6.7 Left: Bi-partite neighbouring co-occurrence graph. Right: second-order graph on neighbouring co-occurrences clustered with CW

To test this on a large scale, the second-order similarity graph for the BNC was computed, excluding the most frequent 2,000 words and drawing edges between words only if they shared at least four left and four right common neighbours. The clusters are checked against a lexicon that contains the most frequent tag for each word in the BNC. The largest clusters are presented in Table 6.2, together with the predominant tags in the BNC.

size	tags(count)	sample words
18432	NN(17120), AJ(631)	secret, officials, transport, unemployment, farm, county, wood, procedure, grounds, ...
4916	AJ(4208), V(343)	busy, grey, tiny, thin, sufficient, attractive, vital, ...
4192	V(3784), AJ(286)	filled, revealed, experienced, learned, pushed, occurred, ...
3515	NP(3198), NN(255)	White, Green, Jones, Hill, Brown, Lee, Lewis, Young, ...
2211	NP(1980), NN(174)	Ian, Alan, Martin, Tony, Prince, Chris, Brian, Harry, Andrew, Christ, Steve, ...
1855	NP(1670), NN(148)	Central, Leeds, Manchester, Australia, Yorkshire, Belfast, Glasgow, Middlesbrough, ...

Table 6.2 The largest clusters of tagset 2 for the BNC

In total, CW produced 282 clusters, of which 26 exceed a size of 100. The weighted average of cluster purity (i.e. the number of predominant tags divided by cluster size) was measured at 88.8%, which exceeds significantly the precision of 53% on word type as reported by Schütze [220].

Again, several hundred clusters, mostly of open word classes are obtained. For computing tagset 2, an efficient algorithm like CW is crucial: the graphs as used for the experiments consist typically of 10,000 to 100,000 vertices and about 100,000 to 1 million edges.

6.6 Combination of Tagsets 1 and 2

Now, there are two tagsets of two different, yet overlapping frequency bands. A large portion of these 8,000 words in the overlapping region is present in both tagsets. Again, a graph is constructed, containing the clusters of both tagsets as vertices; weights of edges represent the number of common elements, if at least two elements are shared. Notice that the graph is bipartite.

And again, CW is used to cluster this graph of clusters. This results in fewer clusters than before for the following reason: while the granularities of tagsets 1 and 2 are both high, they capture different aspects as they are obtained from different sources. Vertices of large clusters (which usually consist of open word classes) have many edges to the other partition's vertices, which in turn connect to yet other clusters of the same word class. Eventually, these clusters can be grouped into one.

Clusters that are not included in the graph of clusters are treated differently, depending on their origin: clusters of tagset 1 are added to the result, as they are believed to contain important closed word class groups. Dropouts from tagset 2 are simply left out, as they mostly consist of small, yet semantically motivated word sets. The total loss of words by disregarding these many but small clusters did never exceed 10% in any experiment. Figure 6.8 illustrates this combination process.

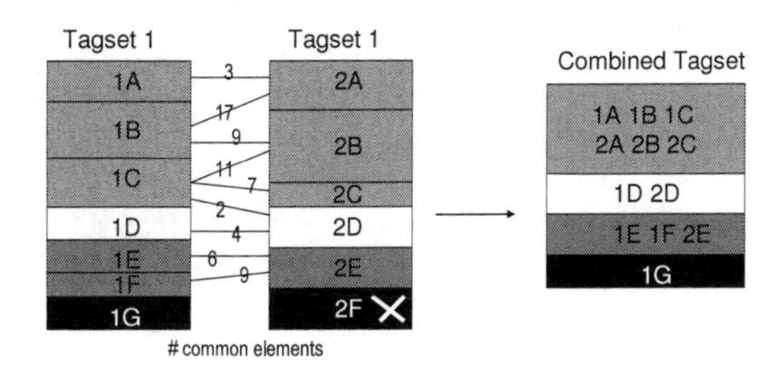

Fig. 6.8 Combination process: tagsets 1 and 2 are related via the number of common elements in their respective clusters. Shades symbolise the outcome of Chinese Whispers on this graph of clusters. Clusters marked with x are not included in the resulting graph of clusters

Conducting this combination yields about 300–600 clusters that will be further used as a lexicon for tagging. As opposed to the observations made in [220], only a handful of clusters are found per open word class, of which most distinctions are syntactically motivated, e.g. adjectives with different case markers. For unsupervised POS tagging, the aim is to arrive at a low number of clusters to mimic the supervised counterparts. A more rigid method to arrive at yet less clusters would be to leave out classes of low corpus frequency.

6.7 Setting up the Tagger

6.7.1 Lexicon Construction

From the merged tagsets, a lexicon is constructed that contains one possible tag (the cluster ID) per word. To increase text coverage, it is possible to include those words that dropped out in the distributional step for tagset 1 into the lexicon. It is assumed that some of these words could not be assigned to any cluster because of ambiguity. From a graph with a lower similarity threshold s (here: such that the graph contains 9,500 target words), neighbourhoods of these words are obtained one at a time. This is comparable to the methodology in [91], where only some vertices are used for clustering and the rest is assimilated. Here, the added target words are not assigned to only one cluster: the tags of their neighbours — if known — provide a distribution of possible tags for these words. Figure 6.9 gives an example: the name 'Miles' (frequency rank 8,297 in the BNC) is rated 65% as belonging to a first name cluster and 35% as last name.

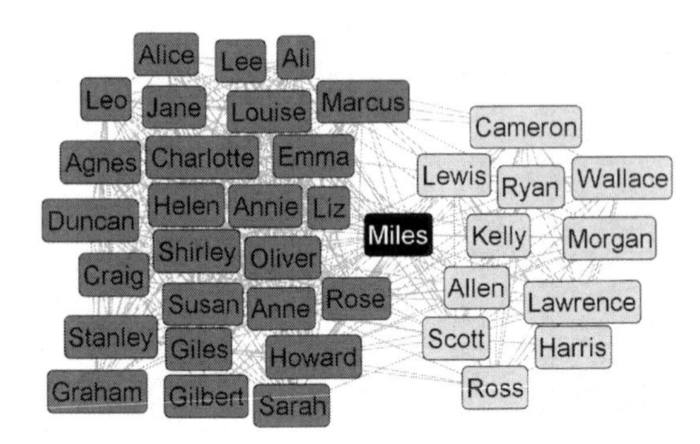

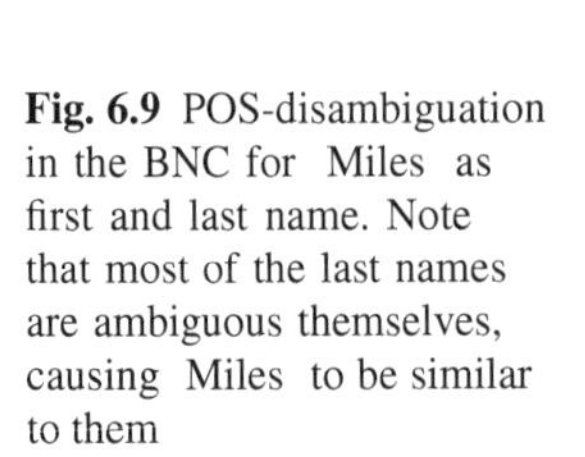
Fig. 6.9 POS-disambiguation in the BNC for Miles as first and last name. Note that most of the last names are ambiguous themselves, causing Miles to be similar to them

6.7.2 Constructing the Tagger

Unlike in supervised scenarios, the task is not to train a tagger model from a small corpus of hand-tagged data, but from the clusters of derived syntactic categories and a large, yet unlabelled corpus. This realises a class-based N-gram model [49].

Here, a simple trigram Viterbi model without re-estimation techniques (such as Baum-Welch) is employed in order not to blur the quality of lexicon construction and to speed up processing. Adapting a previous standard POS-tagging framework [cf. 57], the probability of the joint occurrence of tokens t_i and categories c_i for a sequence of length n is maximised:

$$P(T,C) = \prod_{i=1}^{n} P(c_i|c_{i-1},c_{i-2})P(c_i|t_i) \tag{6.4}$$

The transition probability $P(c_i|c_{i-1},c_{i-2})$ is estimated from word trigrams in the corpus whose elements are all present in the lexicon. The last term of the product, namely $P(c_i|t_i)$, is dependent on the lexicon. If the lexicon does not contain t_i, then c_i only depends on neighbouring categories, i.e. $P(c_i|t_i) = 1$. Words like these are called out-of-vocabulary (OOV) words.

Although Charniak et al. [57] report that using $P(t_i|c_i)$ for the last term leads to superior results in the supervised setting, this 'direct' lexicon probability is used here. The experiments in [57] were carried out for small, labelled corpora in a supervised setting. The main advantage of the current standard model, better smoothing capability, is not an issue when using much larger corpora, as conducted here. For an efficient implementation, beam search (cf. [45] in a tagging context) is employed to keep only the 5 most probable states per token position. The beam width of 5 is a safe choice, as preliminary experiments showed that already a beam width of 3 produces practically equal tagging results compared to using all states.

6.7.3 Morphological Extension

The main performance flaw of supervised POS taggers originates from OOV words. Morphologically motivated add-ons are used e.g. in [66; 104] to guess a more appropriate category distribution based on a word's suffix or its capitalisation. Here, Compact Patricia Trie classifiers (CPT, [see 143]) trained on prefixes and suffixes are employed. For OOV words, the category-wise product of both classifier's distributions serve as probabilities $P(c_i|t_i)$: Let $w = ab = cd$ be a word, a be the longest common prefix of w and any lexicon word, and d be the longest common suffix of w and any lexicon words. Then

$$P(c_i|w) = \frac{|\{u|u = ax \wedge class(u) = c_i\}|}{|\{u|u = ax\}|} \times \frac{|\{v|v = yd \wedge class(v) = c_i\}|}{|\{v|v = yd\}|}. \tag{6.5}$$

CPTs do not only serve as a substitute lexicon component, they also handle capitalisation, camelCase and suffix endings without having to define features explicitly or setting length or maximal number thresholds (as in [104] for suffixes). A similar technique is employed by Cucerzan and Yarowsky [73] in the context of named entity recognition. The author's implementation of CPTs[3] is further used in supervised settings by Witschel and Biemann [248] for compound splitting and in [87] for base form reduction, where it is described in more detail.

6.8 Direct Evaluation of Tagging

Adopting the methodology of Freitag [104], the cluster-conditional tag perplexity PP as the average amount of uncertainty to predict the tags of a POS-tagged corpus, given the tagging with classes from the unsupervised method is measured: for the same corpus tagged with two methods, the measure indicates how well one tagging can be reproduced from the other. Let

$$I_x = -\sum_x P(x) ln P(x) \tag{6.6}$$

be the entropy of a random variable X and

$$M_{XY} = \sum_{xy} P(x,y) ln \frac{P(x,y)}{P(x)P(y)} \tag{6.7}$$

be the mutual information between two random variables X and Y. Then the cluster-conditional tag perplexity for a gold-standard tagging T and a tagging resulting from clusters C is computed as

[3] available as part of the ASV toolbox,
http://wortschatz.uni-leipzig.de/~cbiemann/software/toolbox/index.htm [August 2011].

$$PP = exp(I_{T|C}) = exp(I_T - M_{TC}). \tag{6.8}$$

Minimum PP is 1.0, connoting a perfect congruence with gold standard tags. Below, PP on lexicon words and OOV words is reported separately. The objective is to minimise the total PP.

Unsupervised POS-tagging is meant for yet untagged text, so a system should be robustly performing on a variety of typologically different languages. For evaluating tagging performance, three corpora are chosen: the BNC for English, a 10 million sentences newspaper corpus from LCC for German, and 3 million sentences from LCC's Finnish[4] web corpus. Table 6.3 summarises some characteristics.

Table 6.3 Characteristics of corpora for POS induction evaluation: number of sentences, number of tokens, tagger and tagset size, corpus coverage of top 200 and 10,000 words

lang.	sent.	tokens	tagger	tags	200 cov.	10K cov.
ENG	6M	100M	BNC	84	55%	90%
FIN	3M	43M	Connexor	31	30%	60%
GER	10M	177M	[215]	54	49%	78%

Since a high text coverage is reached with only a few words in English, a strategy that assigns only the most frequent words to sensible clusters already ensures satisfactory performance. In the Finnish case, a high OOV rate can be expected, hampering performance of strategies that cannot cope well with low frequency or unseen words.

To put the results in perspective, the following baselines on random samples of the same 1,000 randomly chosen sentences used for evaluation were computed:

- 1: the trivial top clustering: all words are in the same cluster
- 200: the most frequent 199 words form clusters of their own; all the rest is put into one cluster.
- 400: same as 200, but with 399 most frequent words

Table 6.4 summarises the baselines in terms of PP, along with analogously defined baselines that are needed later for comparison.

6.8.1 Influence of System Components

The quality of the resulting taggers for combinations of several sub-steps is measured using:

- O: tagset 1
- M: the CPT morphology extension

[4] Thanks goes to Connexor Oy, Helsinki, for an academic licence of their Finnish MBT tagger.

Table 6.4 Baselines for various tagset sizes

	English				
baseline	1	200	345	400	619
PP	29.3	3.69	3.17	3.03	2.53
	Finnish				
baseline	1	200	400	466	625
PP	20.2	6.14	5.58	5.46	5.23
	German				
baseline	1	200	400	440	781
PP	18.24	3.32	2.79	2.73	2.46

- T: merged tagsets 1 and 2
- A: adding ambiguous words to the lexicon

Figure 6.10 illustrates the influence of the similarity threshold s for O, O+M and O+M+A for German — for other languages, results look qualitatively similar. Varying s influences coverage on the 10,000 target words. When clustering on very few words, tagging performance on these words reaches a PP as low as 1.25 but the high OOV rate impairs the total performance. Clustering too many words results in deterioration of results — most words end up in one big part. In the medium ranges, higher coverage and lower known PP compensate each other, optimal total *PP*s were observed at target word coverages of 4,000-8,000. The system's performance is stable with respect to changing thresholds, as long as it is set in reasonable ranges. Adding ambiguous words results in a worse performance on lexicon words, yet improves overall performance, especially for high thresholds.

For all further experiments, the threshold s was fixed in a way that tagset 1 consisted of 5,000 words, so only half of the top 10,000 words are considered unambiguous. At this value, the best performance throughout all corpora tested was achieved.

Table 6.5 Results in PP for English, Finnish, German. OOV% is the fraction of non-lexicon words in terms of tokens

lang	words	O	O+M	O+M+A	T+M	T+M+A
EN	total	2.66	2.43	2.08	2.27	2.05
	lex	1.25		1.51	1.58	1.83
	OOV	6.74	6.70	5.82	9.89	7.64
	OOV%	28.07		14.25	14.98	4.62
	tags	619			345	
FI	total	4.91	3.96	3.79	3.36	3.22
	lex	1.60		2.04	1.99	2.29
	OOV	8.58	7.90	7.05	7.54	6.94
	OOV%	47.52		36.31	32.01	23.80
	tags	625			466	
GER	total	2.53	2.18	1.98	1.84	1.79
	lex	1.32		1.43	1.51	1.57
	OOV	3.71	3.12	2.73	2.97	2.57
	OOV%	31.34		23.60	19.12	13.80
	tags	781			440	

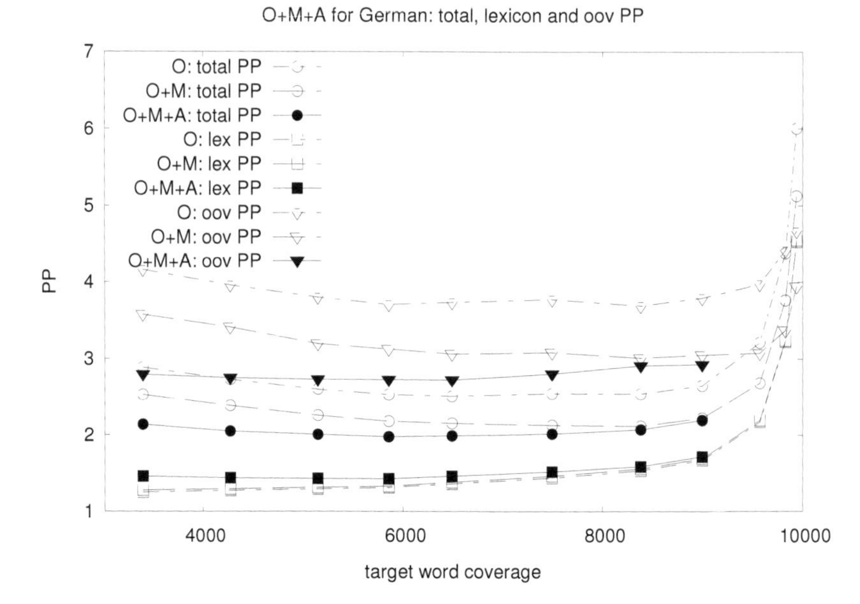

Fig. 6.10 Influence of threshold s on tagger performance: cluster-conditional tag perplexity PP as a function of target word coverage for tagset 1

Overall results are presented in Table 6.5. The combined strategy T+M+A reaches the lowest PP for all languages. The morphology extension (M) always improves the OOV scores. Adding ambiguous words (A) hurts the lexicon performance, but largely reduces the OOV rate, which in turn leads to better overall performance. Combining both partitions (T) does not always decrease the total PP a lot, but lowers the number of tags significantly.

Finnish figures are generally worse than for the other languages, which is consistent with its higher baselines. Differences between languages are most obvious when comparing O+M+A and T+M: whereas for English it pays off much more to add ambiguous words than to merge the two partitions, it is the other way around in the German and Finnish experiments.

To sum up the discussion of results: all introduced steps improve the performance, yet their influence's strength varies. As a sample of the system's output, consider the example in Table 6.6 that has been tagged by the English T+M+A model: as in the example above, 'saw' is disambiguated correctly. Further, the determiner cluster is complete; unfortunately, the pronoun 'I' constitutes a singleton cluster.

The results can be compared to [104]; most other work uses different evaluation techniques that are only indirectly measuring what is tried to optimise here. Unfortunately, Freitag [104] does not provide a total PP score for his 200 tags. He experiments with a hand-tagged, clean English corpus that is not free (the Penn Treebank) and was therefore not an option here. Freitag reports a PP for known

Table 6.6 Tagging example

Word	cluster ID	cluster members (size)
I	166	I (1)
saw	2	*past tense verbs* (3818)
the	73	a, an, the (3)
man	1	*nouns* (17418)
with	13	*prepositions* (143)
a	73	a, an, the (3)
saw	1	*nouns* (17418)
.	116	. ! ? (3)

words of 1.57 for the top 5,000 words (91% corpus coverage, baseline 1 at 23.6), a PP for unknown words without morphological extension of 4.8. Using morphological features the unknown PP score is lowered to 4.0. When augmenting the lexicon with low frequency words via their distributional characteristics, a PP as low as 2.9 is obtained for the remaining 9% of tokens. His methodology, however, does not allow for class ambiguity in the lexicon, the low number of OOV words is handled by a Hidden Markov Model trained with Baum-Welch-Reestimation.

Due to different evaluation tagsets and a different baseline, it is hard to assess whether Freitag's method performs better or worse on information-theoretic measures on English. For most other languages with flatter frequency distributions, Freitag's method can be expected to perform worse because of higher OOV rates resulting from the 5,000 word limit.

6.8.2 Influence of Parameters

A number of parameters for the process of unsupervised POS tagging were introduced at the points where they wee first mentioned. Now, all parameters are listed for recapitulating the possibilities to fine-tune the method. Table 6.7 gives the parameters, a short explanation, and the default setting used in all experiments. Their influence and interplay is outlined as follows. *FEAT* did not show to have a large influence in ranges 100–300. It might be adviseable to use higher values for languages with low Zipfian exponents (such as Finnish) to gain higher text coverage for building tagset 1. When processing small corpora, *TARG* should not be too high, because a low corpus frequency for target words results in unreliable context statistics. The parameter *CWTARG* must be set smaller than *TARG*, Figure 6.10 indicates that 40%–80% of *TARG* is a sensible range. Higher settings result in more words that can overlap for combining the two tagsets.

NB_SIG and *NB_THRESH* go hand in hand to regulate the density of the graph for tagset 2: Lower significance thresholds lead to more edges in the neighbouring co-occurrence graph, higher values for *NB_THRESH* prune edges in the graph for tagset 2. The maximum number of neighbouring co-occurrences per word *NB_MAX* influences the density of the graph for tagset 2, lower settings result in less edges per word. All experiments were carried out with the default value, however, higher

Parameter	Default	Explanation
FEAT	200	Number of feature words for tagset 1 similarities
TARG	10,000	Number of target words for tagset 1 similarities
CWTARG	5000	Number of words that are clustered for tagset 1 amongst *TARG* words, by applying an appropriate similarity threshold s on the graph
TOPADD	9500	Number of words that are considered for adding ambiguous words amonst *TARG* words, by applying an appropriate similarity threshold s on the graph
NB_SIG	2.71	Significance threshold for neighbour-based co-occurrences
NB_THRESH	4	Minimum number of common neighbour-based co-occurrences per side for constituting an edge in the graph for tagset 2
NB_MAX	200	Maximum neighbouring co-occurrences per word to consider for the second-order graph of tagset 2
CONF_OVERLAP	2	Minimum number of common words for connecting partitions in the graph of clusters to merge tagset 1 and 2
BEHEAD	2000	Minimum rank of words to enter the graph for tagset 2
SING_ADD	200	Maximum frequency rank of words to add as singletons, if not already contained in the combined tagset.

Table 6.7 Parameters, default settings and explanation

values lead to more coarse-grained tagsets that e.g. join common and proper nouns. Different settings could prove advantageous for different applications, but no experiments were conducted to measure to what extent.

BEHEAD should be set in a way that stop words are excluded from tagset 2, but considerably lower than *TARG*, to enable sufficient overlap between the two tagsets. Less than a value of 2 for *CONF_OVERLAP* can result in spurious combinations in the graph of clusters, higher values reduce the lexicon size since clusters from tagset 2 are more likely to be excluded.

Adding more singletons amongst the *SING_ADD* most frequent words increases the number of tags, but also the number of trigrams available for training the Viterbi tagger.

A sensible extension would be to limit the total number of tags by excluding those clusters from the combined tagset that have the lowest corpus frequency, i.e. the sum of frequencies of the lexicon words constituting this tag.

6.8.3 Influence of Corpus Size

Having determined a generic setting for the interplay of the different system components (M+T+A), the influence of corpus size on the cluster-conditional tag perplexity PP shall be examined now. For this purpose, taggers were induced from German corpora of varying size from 100,000 sentences up to 40 million sentences, taken from LCC. Evaluation was carried out by measuring PP between the resulting taggers and the hand-tagged German NEGRA corpus [46], testing on all 20,000 sentences of NEGRA. The evaluation corpus was not part of the corpora used for tagger induction. Figure 6.11 provides total PP, the OOV rate and the lexicon size, dependent on corpus size.

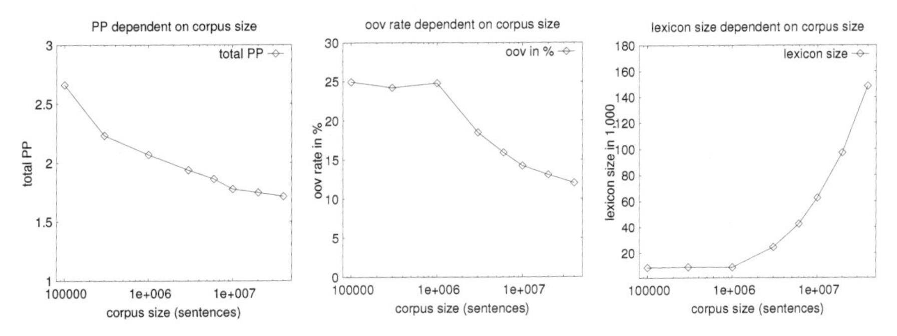

Fig. 6.11 PP, OOV rate and lexicon size vs. corpus size for German

Not surprisingly, a larger corpus for tagger induction leads to better performance levels in terms of PP. The more data provided, the more reliable the statistics for tagset 1, which is reflected in tremendous PP improvement from using 100,000 to 1,000,000 sentences. In this range, tagset 2 is almost empty and does not contribute to the lexicon size, which is mirrored in a constant OOV rate for this range. Above 1 million sentences, the size of tagset 2 increases, resulting in lower PP and OOV rates. The lexicon size explodes to some 100,000 entries for a corpus size of 40 million sentences. Summarising the results obtained by training the unsupervised POS tagger on corpora of various sizes, there always seems to be room for improvements by simply adding more data. However, improvements beyond 10 million sentences are small in terms of PP.

The interpretation of the PP measure is difficult, as it largely depends on the gold standard. While it is possible to relatively compare the performance of different components of a system or different systems along these lines, it only gives a poor impression on the utility of the unsupervised tagger's output. Therefore, several application-based evaluations are undertaken in Section 6.9.

6.8.4 Domain Shifting

Now some light is shed on the advantage of using unsupervised POS tags for domain adaptation. When cultivating an NLP system for one domain or genre and then applying it to a different domain, a major drop in performance can be expected due to different vocabulary and different constructions, cf. [76].

It would be desirable to train on one domain and test performance on the other, using various ways to incorporate the unsupervised POS tags, e.g. inducing separate models or a single model on a mixed corpus and measure the contribution to a task. Apart from corpora from different domains, this would require gold standard data for the same task in different domains, which were unfortunately not available. Reverting to quantitative observations, accompanied by exemplifying data, a general-domain corpus of British English (the BNC as used above) is contrasted with a specialised medical domain corpus of mixed-spelling English (the 2004 MeSH abstracts[5], henceforth called MEDLINE.).

In Table 6.8, OOV rates for two unsupervised POS models trained on these corpora are given. Not surprisingly, a shift in domain results in higher OOV rates: applying a model trained on the specialised domain MEDLINE corpus to the general-domain BNC leads to almost twice the OOV rate. Applying the general domain BNC model to the specialised domain — a more common scenario in domain adaptation — results in a more than 3-fold increase of the OOV rate from 5% to 18.8%.

	BNC model	BNC top 10K	MEDLINE model	MEDLINE top 10K
BNC OOV	7.1%	8.6%	12.0%	20.7%
MEDLINE OOV	18.8%	21.9%	5.0%	9.5%

Table 6.8 OOV rates for unsupervised POS models for BNC and MEDLINE, both in-domain and cross-domain. For reference, also OOV rates with respect to the most frequent 10K words per corpus are given.

While OOV rates only reveal how much vocabulary is covered when shifting domain, it does not reveal the utility of the tagset. Tagsets for specialised corpora often reflect their domain: in addition to core word classes, additional word classes are discovered that could help for certain applications. For example, the tag for units in Table 6.9 can help to project from general-domain units like *kg* or *gallons* to specialised units like *kcal/mol*. The tag for cell lines or viruses might facilitate Information Extraction tasks.

6.8.5 Comparison with Clark [66]

This section aims at comparing the unsupervised tagger to the unsupervised word clustering described in [66] [6].

Clark's system consists of a clustering that is inspired by the well-known k-means algorithm. All words above a certain frequency threshold t are clustered into k clusters by iteratively improving the likelihood of a given clustering. The likelihood is measured with respect to a combination of a bigram class model [183] for distributional evidence, a letter Hidden Markov Model for modeling morphological

[5] http://www.nlm.nih.gov/mesh/filelist.html [Jan 2010]

[6] Thanks goes to Alex Clark for making his clustering software available for download http://www.cs.rhul.ac.uk/home/alexc/pos2.tar.gz [Jan 2010].

rank	size	description	sample words
10	1707	cell lines	F98, ANA-1, HUC, NTERA2, Caco-2BBe, AT5BIVA, YH, TIG-1, EG2+, LNCaP-FGC, IEC-6, Raw264.7, spleen, RKO, H292, HT29, BCE, SRG, MLE-15, S16, Mer+, SP-ir, C-21, SW1990, Caki-2, HT-1080, HT29-Cl.16E, SK-N-SH, MH-S, Haller, MES-SA, CA46, NFS-60, MN9D, MCTC, ...
19	705	viruses	APEC, SIVsm, salmonella, herpes, IKC, virus-2, dengue, Hib, Adv, Pnc, pneumococcal, TMEV, anthrax, THO, BVDV-1, NDV, dengue-2, TCLA, HGV, SV-40, toxoplasma, heartworm, WNV, YF, diphtheria, Ara-, IFV, MLV, cryptococcus, ...
27	474	units	kg, g/l, Hz, hm2, mo, CFU, mm2/s, mm2, dL, hrs, U/ml, min)-1, microV, IU/ml, Pounds, kcal/mol, cm3/min, g/min, microg/ml, metre, PDLs, ml/g, centimetres, gallons, euros, mumol/l, pg, months, nanometres, pmoles, MPa, cm2, MJ/d, bp, francs, IU/L, U/m2/day, g/d, g/kg/min, mol, cfu/cm2, ...
39	213	treatment time adjectives	postsurgery, post-transplant, postload, postfracture, gestation, ago, 12-months, post-release, post-test, postop, postcoitus, postchallenge, posttrauma, postdose, post-insult, postadmission, poststroke, postburn, postresuscitation, post-stress, postpartal, post-challenge, regraft, postnatal, EGA, postinsemination, postaxotomy, post-hatching, posthemorrhage, postmenstrual, ...

Table 6.9 Selected clusters from the MEDLINE clustering (final tagset) with randomly selected words. In total, 479 clusters were obtained, which were ordered by decreasing size to rank them. These clusters reflect domain specific word classes that are usually not found in general-domain corpora.

similarities and a frequency prior. The parameter k can be used to set the desired granularity of the clustering.

In contrast to the method described in this paper, all words of a given corpus are clustered: words with frequency of t or less are simply clustered together in one large cluster. This means that there are no OOV words in the corpus w.r.t. the clustering. The method as it stands does not allow assigning tags for unknown words (in new text) from the context alone. Also, the same word always gets assigned the same tag, there is no mechanism that accounts for ambiguity.

Due to the computational cost of the clustering, which increases for larger k, the methods would not run on corpora larger than the BNC, which took up to a week in CPU time[7] for large k. In contrast, the method presented here induces a model for the BNC in a few hours. The default settings (t=5, 10 iterations) were used to produce clusterings for varying k for the BNC and a German corpus of 5M sentences, assigning cluster IDs as tags for the same evaluation corpus used in Section 6.8 above.

The V-measure [205] is the harmonic mean of two measures: Homogeneity h and Completeness c. h quantifies how well the system assignment can be mapped to the gold standard, not penalising too fine-grained system clusters. This measure is very closely related to EP as defined above. The symmetrically defined c quantifies how well the gold standard can be mapped to the system assignment, at this penalising

[7] on a decent PC in 2009

fine-grained system distinctions. Both h and c are normalised and take on values between 0 (undesirable) and 1 (desirable), serving as analogies to Precision and Recall.

Table 6.10 shows the V-measure for the German and English data for different k and different baselines in comparison with the M+T+A models described above.

German

System	Completeness	Homogeneity	V-measure
base-1	1	0	0
base-200	0.6096	0.58626	0.5977
base-440	0.5798	0.6536	0.6145
base-781	0.5539	0.6878	0.6136
Clark-128	0.5481	0.8407	0.6636
Clark-256	0.5221	0.8941	0.6592
Clark-440	0.4982	0.9079	0.6434
unsupos-440	0.5670	0.8604	**0.6835**

English BNC

System	Completeness	Homogeneity	V-measure
base-1	1	0	0
base-200	0.6988	0.6172	0.6555
base-345	0.6697	0.6612	0.6654
base-400	0.6633	0.6739	0.6686
Clark-128	0.6228	0.8204	**0.7081**
Clark-256	0.5841	0.8554	0.6942
Clark-345	0.5727	0.8708	0.6910
unsupos-345	0.6349	0.766	0.6943

Table 6.10 V-measure evaluation for German and English: Baselines, Clark's system (with number of clusters) and this system for the German 40M sentence corpus and the BNC.

From Table 6.10, it becomes clear that the V-measure is relatively robust across a wide range of the number of clusters. Baseline scores demonstrate the tradeoff between h and c. Clark's system produces higher h and lower c than the system presented here for the same number of clusters. This points at a different cluster size distribution: Clark's cluster size distribution on the token level is flatter. For German, the system presented here outperforms Clark's system on the V-measure for all k tested. For English, Clark's system with 128 clusters shows a higher performance. Overall, the systems exhibit a similar performance, with the system presented here being more expressive (with regard to ambiguous words) and more flexible (with respect to unseen types), yet having a higher OOV rate.

6.9 Application-based Evaluation

POS-taggers are a standard component in any applied NLP system. In this section, a number of NLP tasks are cast as machine learning problems: the POS-tagger component provides some of the features that are used to learn a function that assigns a label to unseen examples, characterised by the same set of features as the examples used for training. In this setting, it is straightforward to evaluate the contribution of POS-taggers — be they supervised or unsupervised — by providing the different POS-tagging annotations to the learning algorithm or not.

Having advanced machine learning algorithms at hand that automatically perform feature weighting and selection, the standard approach to NLP systems is to try everything but the kitchen sink and to leave the choice of discriminating between useful and unhelpful features to the learning algorithm. In Structure Discovery, the aim is to use as many unsupervised and knowledge-free methods as possible for providing these features. This section gives examples on how to do this for POS-like annotations.

The task-performing systems for application-based evaluation were chosen to cover two different machine learning paradigms: kernel methods in a word sense disambiguation (WSD) system and Conditional Random Fields (CRFs, see [148]) for supervised POS, named entity recognition (NER) and chunking. Some results of this section have been previously published in [32].

All evaluation results are compared in a pair-wise fashion using the approximate randomisation procedure of [185] as significance test. A significant difference with $p<0.01$ here means that the test is more than 99% sure that the difference has not been caused by chance.

6.9.1 Unsupervised POS for Supervised POS

It might seem contradictory to evaluate an unsupervised POS tagger in terms of the contribution it can make to supervised POS tagging. While there exist high precision supervised POS taggers and elaborate feature sets have been worked out, it does not seem necessary to create an unsupervised tagger in presence of training data. This, however, changes if one looks at different domains or languages. In these settings, any method that can help reduce the amount of training data is a contribution to development speed and cost of natural language processing systems. In recent years, distributional features found their way into supervised POS taggers even for general-domain English.

This section examines the contribution of the unsupervised tagger as a feature in supervised POS tagging. It is shown that the unsupervised tags capture structural regularities beyond standard features such as capitalisation and affixes.

Here, first-order Conditional Random Fields are used for supervised machine learning, leveraging the CRF++[8] implementation. In order to test the contribution of unsupervised POS features, four systems with different feature combinations are compared in terms of performance:

- System A: Only lexical features, time-shifted by -2, -1, 0, 1, 2
- System B: Like System A. Additionally, features time-shifted by -1, 0, 1 for capitalisation, number, 2-letter prefix and 2-letter suffix
- System C: Like System A. Additionally, unsupervised labels time-shifted by -2, -1, 0, 1, 2 as assigned by the tagger model induced on 40 million sentences from the Wortschatz project as evaluated in Figure 6.11,
- System D: Combination of all features present in systems A, B and C.

Training sets of varying sizes are selected randomly from all 20,000 sentences of the hand-tagged NEGRA corpus for German, the respective remainders are used for evaluation. Results are reported in tagging accuracy (number of correctly assigned tags divided by total number of tokens), averaged over three different splits per training size each. Figure 6.12 shows the learning curve.

Results indicate that supervised tagging can clearly benefit from unsupervised tags: between 30% and 50% training with unsupervised tags, the performance on 90% training without the unsupervised extension is surpassed comparing systems D and B. At 90% training, error rate reduction of system D over B is 24.3%, indicating that the unsupervised tagger grasps very well the linguistically motivated syntactic categories and provides a valuable feature to either reduce the size of the required annotated training corpus or to improve overall accuracy. Comparing the gains of systems B, C and D over system A, it is possible to conclude that the unsupervised features provide more than simple capitalisation, number or affix features, since their combination D significantly outperforms systems B and C.

Probably also due to a more advanced machine learning paradigm, system D with its 0.9733 accuracy compares favourably to the performance of [45], who reports an accuracy of 0.967 at 90% training on the same data set — eqal to the performance of system C. As far as known to the author, System D constitutes state-of-the art for German POS tagging. Similarly, the currently best published result on Penn Treebank POS tagging for English uses unsupervised features from the system described here, see [227].

When swapping the unsupervised POS features in system D with the 128 clusters from Clark's method — the k for which the best V-measure were obtained in Section 6.8.5 — equal performance of 0.9733 on the same splits for 90% training was measured. Combining unsupervised features and Clark's features, precision is further improved to 0.9743. Since the improvements of the single unsupervised features do not fully wipe out each other, it can be concluded that the two clustering methods capture slightly different aspects of syntactic similarity.

[8] available at http://crfpp.sourceforge.net/ [version 0.53].

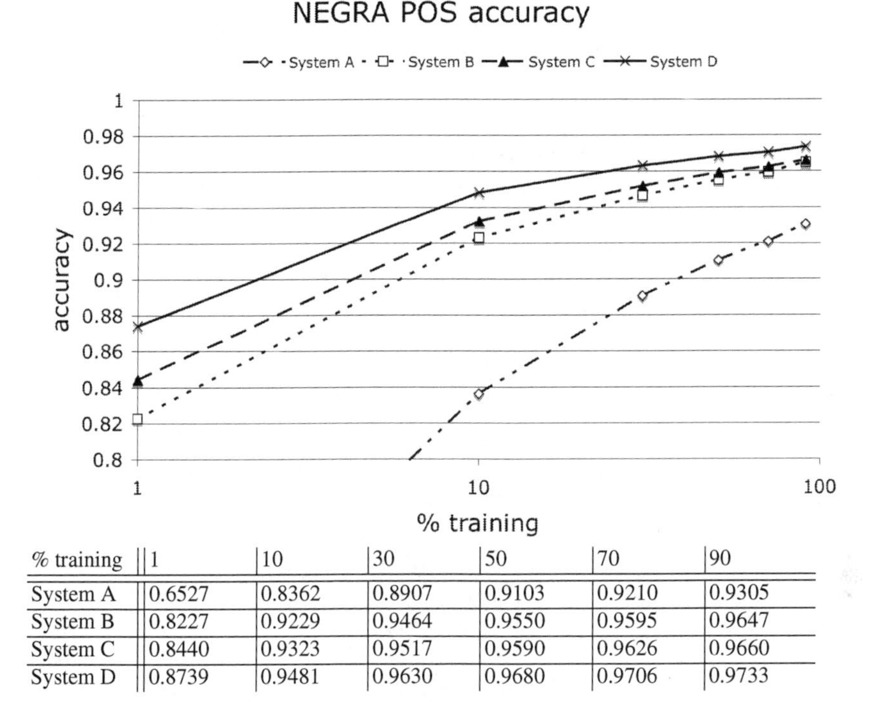

% training	1	10	30	50	70	90
System A	0.6527	0.8362	0.8907	0.9103	0.9210	0.9305
System B	0.8227	0.9229	0.9464	0.9550	0.9595	0.9647
System C	0.8440	0.9323	0.9517	0.9590	0.9626	0.9660
System D	0.8739	0.9481	0.9630	0.9680	0.9706	0.9733

Fig. 6.12 Learning curve for supervised POS-tagging with and without using unsupervised POS features (accuracy).

6.9.2 Unsupervised POS for Word Sense Disambiguation

The task in word sense disambiguation (WSD) is to assign the correct word sense to ambiguous words in a text based on the context. The senses are provided by a sense inventory (usually WordNet, [178]). Supervised WSD is trained on examples where the correct sense is provided manually, and tested by comparing the system's outcome on held-out examples.

In the WSD literature, many algorithms have been proposed, characterised by different feature sets and classification algorithms. The state of the art supervised WSD methodology, reporting the best results in most of the Senseval-3 lexical sample tasks [171] in different languages, is based on a combination of syntagmatic and domain kernels [112] in a Support Vector Machine classification framework. For more on WSD, see Chapter 7.

A great advantage of the methodology of [111; 112] is that all its pre-processing steps are also unsupervised and knowledge-free and therefore comply to the SD paradigm. It is shown here that the only language-dependent component in the system of [111] — a supervised POS-tagger — can safely be replaced by the unsupervised POS-tagger.

Kernel WSD basically encompasses two different aspects of similarity: domain aspects, mainly related to the topic (i.e. the global context) of the texts in which the word occurs, and syntagmatic aspects, concerning the lexical-syntactic pattern in the local contexts. Domain aspects are captured by the domain kernel, while syntagmatic aspects are taken into account by the syntagmatic kernel.

The domain kernel handles domain aspects of similarity among two texts based on the Domain Model as introduced in [112], which is a soft clustering of terms reflecting semantic domains. On the other hand, syntagmatic aspects are probably the most important evidence while recognising sense similarity. In general, the strategy adapted to model syntagmatic relations in WSD is to provide bigrams and trigrams of collocated words as features to describe local contexts [250]. The main drawback of this approach is that non-contiguous or shifted collocations cannot be identified, decreasing the generalisation power of the learning algorithm.

The syntagmatic kernel allows estimating the number of common non-continuous subsequences of lemmas (i.e. collocations) between two examples, in order to capture syntagmatic similarity. Analogously, the POS kernel is defined to operate on sequences of parts-of-speech. The syntagmatic kernel is given by a linear combination of the collocation kernel and the POS kernel.

The modularity of the kernel approach makes it possible to easily compare systems with different configurations by testing various kernel combinations. To examine the influence of POS-tags, two comparative experiments were undertaken. The first experiment uses only the POS kernel, i.e. the POS labels are the only feature visible to the learning and classification algorithm. In a second experiment, the full system as in [111] is tested against replacing the original POS kernel with the unsupervised POS kernel and omitting the POS kernel completely. Table 6.11 summarises the results in terms of Senseval scores for WSD, tested on the lexical sample task for English. The unsupervised POS annotation was created using the BNC tagger model, see Section 6.8.

Table 6.11 Comparative evaluation on Senseval scores for WSD. Differences within systems are not significant at p<0.1

System	only POS	full
no POS	N/A	0.717
supervised POS	0.629	0.733
unsupervised POS	0.633	0.735

Results show that POS information is contributing only to a small extent to WSD accuracy in the full WSD system. Using the unsupervised POS-tagger results in a slight performance increase, improving over the state of the art in this task, which has been achieved with the same system using supervised POS-tags. In conclusion, supervised tagging can safely be exchanged in kernel WSD with the unsupervised variant. Replacing the only pre-processing step that is dependent on manual resources in the system of [111], state-of-the-art supervised WSD is proven to not being dependent on any linguistic pre-processing at all.

Gains in using an unsupervised tagger for WSD can probably be attributed to the finer distinctions the unsupervised tagger makes. E.g. a separate tag for professions can help to generalise over this category. While it is arguable whether this distinction should be part of a standard POS tagset, since this is rather a semantic than a syntactic restriction, it is desirable from the point of view of this application.

6.9.3 Unsupervised POS for NER and Chunking

Named entity recognition (NER) is the task of finding and classifying named entities, such as persons, organisations and locations. Chunking is concerned with shallow syntactic annotation; here, words in a text are labelled as being syntactically correlated, e.g. in noun phrases, verb phrases and prepositional phrases. For performing NER and chunking, these applications are perceived as a tagging task: in each case, labels from a training set are learned and applied to unseen examples. In the NER task, these labels mark named entities and non-named entities, in the chunking task, the respective phrases or chunks are labelled.

For both tasks, the MALLET tagger [162] is trained. It is based on first-order Conditional Random Fields (CRFs), which define a conditional probability distribution over label sequences given a particular observation sequence. The flexibility of CRFs to include arbitrary, non-independent features makes it easy to supply either standard POS-tags, unsupervised POS-tags or no POS-tags to the system without changing its overall architecture.

The tagger operates on a different set of features for the two tasks. In the NER system, the following features are accessible, time-shifted by -2, -1, 0, 1, 2:

- the word itself
- its POS-tag
- Orthographic predicates
- Character bigram and trigram predicates

In the case of chunking, features are only time-shifted by -1, 0, 1 and consist only of:

- Word itself
- POS-tag

This simple feature set for chunking was chosen to obtain a means of almost direct comparison of the different POS schemes without blurring results by other features or system components. Per system, three experiments were carried out, using standard POS features, unsupervised POS features and no POS features.

To evaluate the performance on NER, the methodology as proposed by the providers of the CoNLL-2002 [206] dataset is adopted: for all settings, the performance in terms of the F1[9] measure is reported. Here, the Dutch dataset is employed,

[9] F1 is the harmonic mean of precision P (number of correct divided by number of assigned labels) and recall R (number of correct divided by number of all labels), $F1 = \frac{2PR}{P+R}$ cf. [239].

the unsupervised POS-tagger is induced on the 70 million token Dutch CLEF corpus, see [192]. Table 6.12 summarises the results of this experiment for selected categories using the full training set for training and evaluating on the test data.

Table 6.12 Comparative evaluation of NER on the Dutch CoNLL-2002 dataset in terms of F1 for PERson, ORGanisation, LOCation, MISCellaneous, ALL. No differences are significant with p<0.1

Category	PER	ORG	LOC	MISC	ALL
no POS	0.8084	0.7445	0.8151	0.7462	0.7781
sup. POS	0.8154	0.7418	0.8156	0.7660	0.7857
unsup. POS	0.8083	0.7357	0.8326	0.7527	0.7817

The scores in Table 6.12 indicate that POS information is hardly contributing anything to the system's performance, be it supervised or unsupervised. This indicates that the training set is large enough to compensate for the lack of generalisation when using no POS-tags, in line with e.g. [15] and [43]. The situation changes when taking a closer look on the learning curve, produced by using train set fractions of differing size. Figure 6.13 shows the learning curves for the categories LOCATION and the (micro average) F1 evaluated over all the categories (ALL).

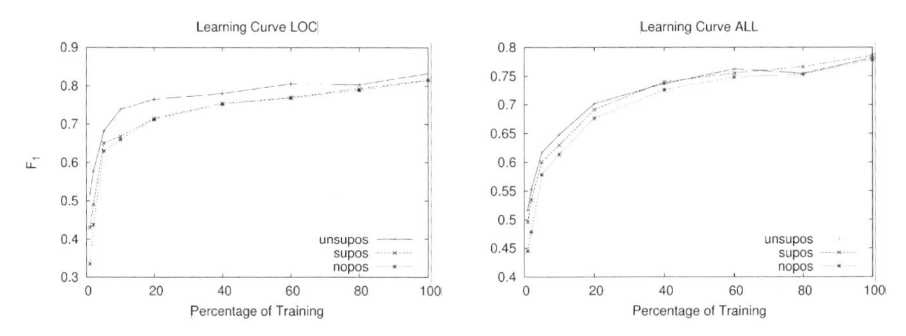

Fig. 6.13 Learning curves in NER task for category LOC and combined category

On the LOCATION category, unsupervised POS-tags provide a high generalisation power for a small number of training samples. This is due to the fact that the induced tagset treats locations as a different tag; the tagger's lexicon plays the role of a gazetteer in this case, comprising 765 lexicon entries for the location tag. On the combination of ALL categories, this effect is smaller, yet the incorporation of POS information outperforms the system without POS for small percentages of training.

This disagrees with the findings of [105], where features produced by distributional clustering were used in a boosting algorithm. Freitag reports improved performance on PERSON and ORGANISATION, but not on LOCATION, as compared to not using a tagger at all.

Experiments on NER reveal that POS information is not making a difference, as long as the training set is large enough. For small training sets, usage of unsupervised POS features results in higher performance than supervised or no POS, which can be attributed to its finer-grained tagset that directly indicates types of named entities.

Performance of the simple chunking system was tested using different portions of the training set as provided in the English CoNLL-2000 data [236] for training, evaluation was carried out on the provided test set. Performance is reported in Figure 6.14.

As POS is the only feature that is used here apart from the word tokens themselves, and chunking reflects syntactic structure, it is not surprising that providing this feature to the system results in increased performance: both kinds of POS significantly outperform not using POS ($p<0.01$). In contrast to the previous systems tested, using the supervised POS labels resulted in significantly better chunking ($p<0.01$) than using the unsupervised labels. This can be attributed to a smaller tagset for supervised POS, providing more reliable statistics because of less sparseness. Further, both supervised tagging and chunking aim at reproducing the same perception of syntax, which does not necessarily fit the distributionally acquired classes of an unsupervised system. Despite the low number of features, the chunking system using supervised tags compares well with the best system in the CoNLL-2000 evaluation (F1=0.9348).

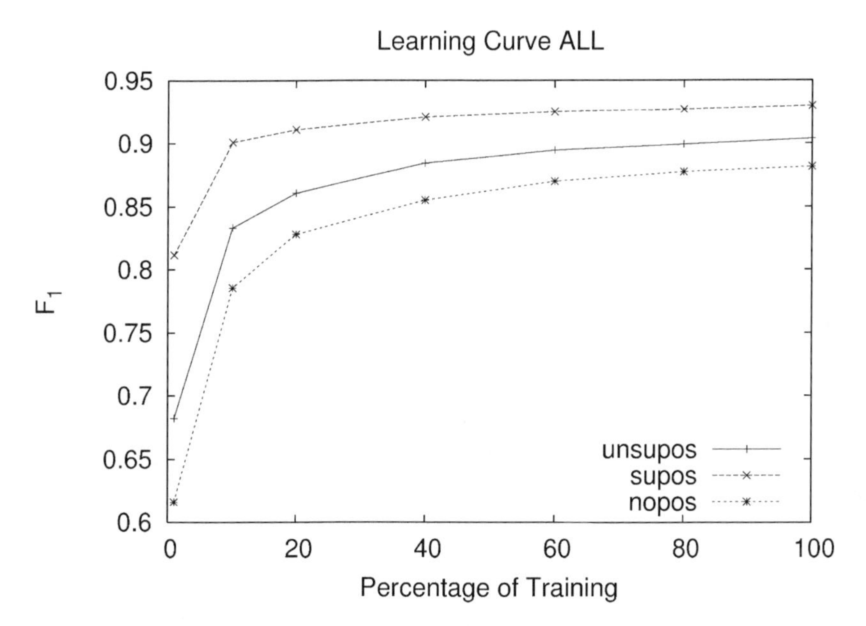

Fig. 6.14 Learning curve for the chunking task in terms of F1. Performance at 100% training is 0.882 (no POS), 0.904 (unsupervised POS) and 0.930 (supervised POS), respectively

6.10 Conclusion on Unsupervised POS Tagging

An unsupervised POS-tagging system was described in detail and evaluated directly and indirectly on various languages and tasks. In difference to previous approaches to unsupervised POS-tagging, this method allows for a larger lexicon, where also POS ambiguities are handled. Further, the discovery of the number of POS categories is part of the method, rather than chosen beforehand.

Comparison with another unsupervised word clustering method shows that the model presented here differs from the system presented in [66]. Both systems yield competitive scores on evaluations, while the system described here is more expressive and faster to induce. The takeaway, however, is that combining several different unsupervised systems as features empowers supervised systems to reach higher performance levels in tasks.

Evaluation on typologically different languages demonstrated the language-independence and robustness of the method. It was demonstrated that for many tasks that use POS as a pre-processing step, there is no significant difference in results between using a trained POS-tagger or the unsupervised tagger presented here. This shows the viability of the Structure Discovery approach.

As far as performance in applications is concerned, the manual efforts necessary to construct a POS-tagger should rather be invested in collecting a large basis of text of the target domain or language, which can be also used for other purposes besides training the unsupervised POS-tagger.

An implementation of the unsupervised POS-tagger system by Andreas Klaus is available for download[10]. This implementation uses the parameter names as given in Table 6.7. On the same page, the tagger models listed in Table 6.13 can be obtained.

Table 6.13 Available taggermodels to date for 14 languages, with corpus size (million sentences), lexicon size (thousand words) and number of tags

Language	Source	Sent.	Lex. Size	Tagset Size
Catalan	LCC	3M	50K	369
Czech	LCC	4M	71K	538
Danish	LCC	3M	43K	376
Dutch	LCC	18M	140K	332
English	BNC	6M	26K	344
English	MEDLINE	34M	118K	479
Finnish	LCC	11M	130K	444
French	LCC	3M	42K	358
German	LCC	40M	258K	395
Hungarian	LCC	18M	180K	332
Icelandic	LCC	14M	132K	326
Italian	LCC	9M	85K	381
Norwegian	LCC	16M	135K	393
Mex. Spanish	LCC	4M	34K	414
Swedish	LCC	3M	43K	370

[10] http://wortschatz.uni-leipzig.de/~cbiemann/software/unsupos.html [July 7th, 2007]

Chapter 7
Word Sense Induction and Disambiguation

Abstract Major difficulties in language processing are caused by the fact that many words are ambiguous, i.e. they have different meanings in different contexts, but are written (or pronounced) in the same way. While syntactic ambiguities have already been addressed in the previous chapter, now the focus is set on the semantic dimension of this problem. In this chapter, the problem of word sense ambiguity is discussed in detail. A Structure Discovery process is set up, which is used as a feature to successfully improve a supervised word sense disambiguation (WSD) system. On this basis, a high-precision system for automatically providing lexical substitutions is constructed.

When approaching ambiguity in the Structure Discovery framework, two steps towards the automatic labelling of homonymous words with their senses can be distinguished and correspond to the two directions in Figure 1.1:

- *Word Sense Induction (WSI)* (also word sense discrimination) is the step of identifying the different word senses or usages from corpus data. This is a clustering task.
- *Word Sense Disambiguation (WSD)* assigns the correct sense from a given set of senses to occurrences of ambiguous words in the text. This is a tagging task.

7.1 Related Work on Word Sense Induction

Both steps of word sense induction and disambiguation have been previously examined extensively. For WSI the general methodology [see e.g. 191; 221; 189; 247; 196; 202; 41] is to cluster the word co-occurrences of a target word to arrive at sets of co-occurrences that represent the different senses, with some variety regarding the context window size, dimensionality reduction techniques and the order of co-occurrences.

WSD has been predominantly performed using a pre-defined sense inventory like WordNet, see [3] and the Senseval/Semeval competitions [e.g. 172; 5; 89] for an overview. Here, the WordNet sense is either assigned using lexical overlap between the dictionary definition and the actual context, or by training a classifier on the hand-labelled SemCor corpus[1].

The study of Agirre et al. [6] proceeds just in the spirit of SD to evaluate the contribution of HyperLex graph clustering ([240], see Section 4.1.3) in the following way: from an unannotated corpus, a sense inventory is induced that is used to assign sense labels for words in a sense-tagged corpus. These are employed to create a mapping between the annotated and the induced senses. A held-out sense-tagged corpus serves as the basis for evaluating the quality of the induced annotation via the mapping and was used to optimise the parameters of HyperLex. Interestingly, the best results were obtained with a high granularity of microsenses, rather than with coarse-grained distinctions that are supported by a broader data basis per sense. This suggests two things: first, the manually assigned senses and the automatically induced senses do not match well; this is why a finer-grained sense induction produces a purer mapping. Second, the data basis for microsenses is sufficently large to assign them with an accuracy that almost reaches state of the art performance for knowledge-based WSD systems despite the lossy mapping. When comparing HyperLex to PageRank [48] for selecting root vertices in [7], very similar results suggest that the key issue in graph-based WSI is not so much the clustering algorithm used, but rather the construction of the underlying graphs.

7.2 Task-oriented Definition of WSD

As an enabling step for semantic applications, Word Sense Disambiguation (WSD) is the task of assigning word senses for ambiguous words in context. In the supervised setting, a sense-labelled training corpus is used to train a model for each target word. This model is used to classify the occurrence of a target word in an unseen context into one of the senses that occurred in the training. In the knowledge-based setting, a semantic resource (like WordNet) is employed to identify senses in context by merely using the semantic resource itself, not the training examples.

In particular for semantic search with matching beyond keywords, one is interested in the possible substitutions for a word in context to be able to expand the index with these. In case of ambiguous words, it is necessary to identify the correct sense first to avoid spurious expansions that lead to mismatches.

A system for lexical substitution thus can be realised by a WSD system that assigns word senses in context and that is equipped with a set of substitutions per sense. Notice that when interested merely in lexical substitution quality, the sense inventory itself plays only an intermediate role as we are not interested in identifying one of possible senses given by the inventory, but in a set of *acceptable substitutions*. In particular, we do not need to map the inventory used for this to other inventories like WordNet. This gives rise to the use of Structure Discovery procedures, since an explicit mapping to such a symbolic resource is not needed for this task.

[1] available at http://multisemcor.itc.it/semcor.php [June 1st, 2007].

High WSD performance scores using WordNet suffer from the extremely fine-grained distinctions that characterise the resource and by the relatively little available data for senses in contexts (cf. e.g. [3]). For example, of the eight noun senses of 'hook', four refer to a bent, curvy object. However, in the entire SemCor [170] there is only one occurrence recorded for this sense altogether, so for most of the senses the only data available are the glosses and the relations to other synsets. Even if some fine-grained classes are combined by clustering WordNet senses [173; 129], alignment of the sense inventory and the target domain or application remains a problem. A further problem with pre-defined word sense inventories is that their senses often do not match the application domain, and there is no consensus on the granularity level of senses, even within single sense inventories. Kilgarriff [139] states "that word senses are only ever defined relative to a set of interests. The set of senses defined by a dictionary may or may not match the set that is relevant for an NLP application". Taking the dissent among lexicographers and the fuzziness of the term 'word sense' into account, Cohn [69] even considers the task of WSD as being ill-defined. Therefore, a domain- and corpus-specific word sense inventory is, besides being the only possibility when working in SD, also highly desired independent of the processing paradigm. This is also reflected in the study of Schütze and Pedersen [217], one of the rare studies where WSD showed to have a positive effect on information retrieval when using senses induced from a corpus: most other studies [241, inter al.] reported negative effects when applying WSD with a predefined sense inventory in a search context.

7.3 Word Sense Induction using Graph Clustering

In this section, an SD approach to word sense induction and disambiguation is described. Similar to the approach in [247], a word graph around each target word is constructed. Here, sentence-based co-occurrence statistics from a large corpus (cf. Section 3.2.1) is used as a basis to construct several word graphs for different parameterisations: since it is not clear what sense granularity is optimal for the application at hand, several levels of granularity are offered to the consuming application. Some of the results presented here have been published in [30].

Significant co-occurrences between all content words (nouns, verbs, adjectives as identified by POS tagging) are computed from a corpus. The full word graph for a target word is defined as the open neighbourhood of the target word vertex. Edges between words that co-occur only once or with significance smaller than 6.63 (1% confidence level) are omitted. Aiming at different granularities of usage clusters, the graph is parameterised by a size parameter t and a density parameter n: Only the most significant t co-occurrences of the target enter the graph as vertices, and an edge between vertices is drawn only if one of the corresponding words is contained in the most significant n co-occurrences of the other.

7.3.1 Graph Clustering Parameterisation

As described in Chapter 4, the neighbourhood graph is clustered with Chinese Whispers. Three ways of vertex weighting (cf. Section 4.2.3) are used: (a) dividing the influence of a vertex in the update step by the degree of the vertex, (b) dividing by the natural logarithm of the degree + 1 and (c) not doing vertex weighting. The more aggressive the downweighting, the higher granularity is expected for the clustering.

Figure 7.1 shows two different sample clusterings for the target 'bank'.

1. Clustering for n=50,t=200, vertex weighting (a)

- bank0: largest, north, branches, eastern, opposite, km, east, west, branch, Thames, banks, located, Danube, town, south, situated, River, Rhine, river, western, commercial, central, southern
- bank1: right, left
- bank2: money, robbers, deposit, robberies, cash, currency, account, deposits, Bank, robbery, funds, financial, banking, loans, notes, robber, rob, accounts, credit, assets, teller, Banco, loan, investment, savings

2. Clustering for n=50, t=100, vertex weighting (c)

- bank0: eastern, banks, central, river, km, western, south, southern, located, largest, east, deposits, commercial, Thames, north, west, Danube, town, situated, Rhine, River
- bank1: branches, branch
- bank2: robberies, robbers, robbery, robber
- bank3: right, left, opposite
- bank4: loans, cash, investment, teller, account, financial, loan, deposit, credit, funds, accounts, assets, savings, banking, money, rob
- bank5: Banco, currency, notes, Bank

Fig. 7.1 Clusterings with different CW parameterisations for the same target word 'bank', showing different levels of granularity.

It can be observed that the clusterings are probably too fine-grained for word sense induction purposes, since the monetary sense is spread over several clusters in the second clustering. Some clusters are related to verb usages of the target, e.g. bank1 from the first clustering. Sometimes, different meanings are also lumped together in a single cluster. However, using the second clustering as a feature enables the system to assign the river bank of the Danube, given e.g. that the river bank of the Thames was found in the training. It is emphasised that no tuning techniques are applied to arrive at the 'best' clustering. Rather, several clusterings of different granularities as features are made available to a supervised system. Note that this is different from [6], where a single global clustering was used directly in a greedy mapping to senses.

7.3.2 Feature Assignment in Context

For a given occurrence of a target word, the overlap in words between the textual context and all clusters from the neighbourhood graph is measured by counting the number of common words in cluster and context. The cluster ID of the cluster with the highest overlap is assigned as a feature. This can be viewed as a word sense induction system in its own right. At this, several clusterings from different parameterisations are used to form distinct features, which enables the machine learning algorithm to pick the most suitable cluster features per target word when building the classification model.

Figure 7.2 exemplifies the interplay of clustering and assigning the cluster ID in context for the noun. It also exemplifies a hierarchical clustering, cf. Section 4.2.7. Cluster names in the figure have been assigned manually for better readability.

7.4 Evaluation of WSI Features in a Supervised WSD System

This section describes experiments with using induced word senses as a feature in a supervised WSD system. After describing the machine learning setup, the system is tested on two different datasets, showing improvements when employing Structure Discovery features.

7.4.1 Machine Learning Setup for Supervised WSD System

Apart from lexical features and part-of-speech (POS) sequences, successful approaches to supervised WSD employ features that aim at modeling the topicality of the context (e.g. LSA [111], LDA [51], see Section 6.9.2): A context is translated into a topic vector, which informs the classifier about the general topic of the context. For example, while a monetary context for 'bill' indicates its sense of 'bank note', a restaurant context would hint at its sense of 'check'. These topical features are computed from a document collection and are the same for all target words. Topic Signatures [159] is an attempt to account for differences in relevant topics per target word. Here, a large number of contexts for a given sense inventory are collected automatically using relations from a semantic resource, sense by sense. The most discriminating content words per sense are used to identify a sense in an unseen context. This approach is amongst the most successful methods in the field. It requires, however, a semantic resource of sufficient detail and size (here: WordNet) and a sense-labelled corpus to estimate priors from the sense distribution.

First, a baseline system for WSD using lexical and part-of-speech (POS) features is described. Then, its augmentation with cluster features from the WSI system described above is laid out.

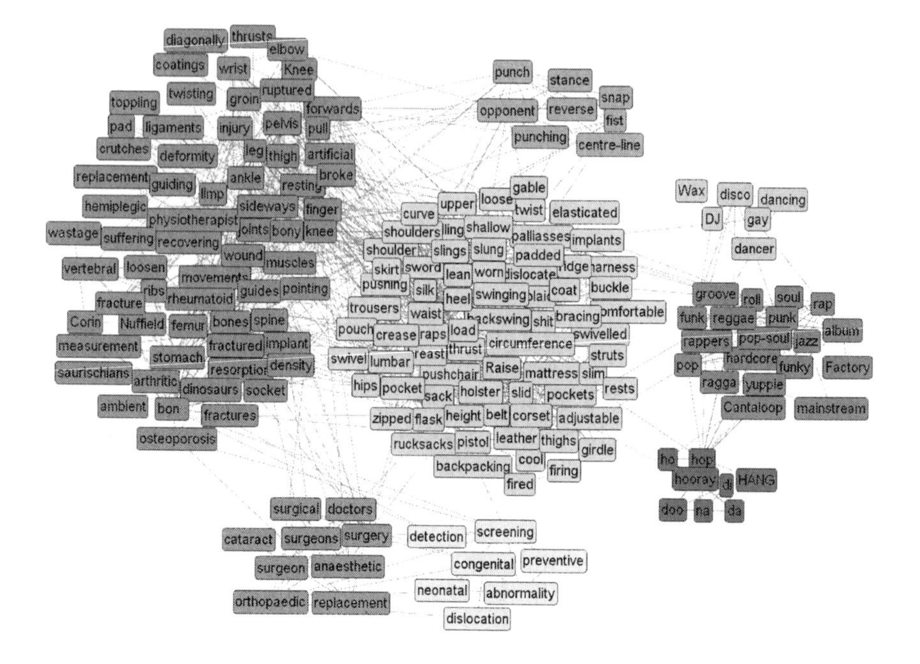

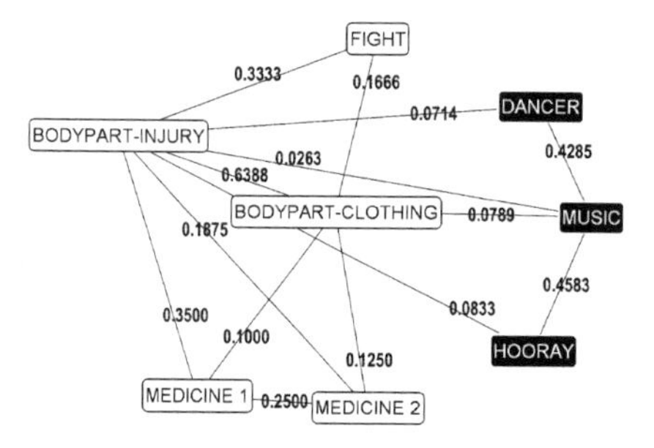

- **FIGHT**: The punching hip , be it the leading hip of a front punch or the trailing hip of a reverse punch , must swivel forwards , so that your centre-line directly faces the opponent .
- **MUSIC**: This hybrid mix of reggae and hip hop follows acid jazz , Belgian New Beat and acid swing the wholly forgettable contribution of Jive Bunny as the sound to set disco feet tapping .
- **DANCER**: Sitting back and taking it all in is another former hip hop dancer , Moet Lo , who lost his Wall Street messenger job when his firm discovered his penchant for the five-finger discount at Polo stores
- **HOORAY**: Ho , hey , ho hi , ho , hey , ho , hip hop hooray , funky , get down , a-boogie , get down .
- **MEDICINE**: We treated orthopaedic screening as a distinct category because some neonatal deformations (such as congenital dislocation of the hip) represent only a predisposition to congenital abnormality , and surgery is avoided by conservative treatment .
- **BODYPART-INJURY**: I had a hip replacement operation on my left side , after which I immediately broke my right leg .
- **BODYPART-CLOTHING**: At his hip he wore a pistol in an ancient leather holster .

Fig. 7.2 WSI Example of 'hip' with hierarchical agglomerative CW and usages. Top: first level clustering. Bottom left: second-level clustering. Bottom right: Sample usages from the BNC

A strong baseline system with standard WSD features was set up to compare against a system augmented with WSI cluster features. The baseline system uses 15 symbolic features per context (number of features in parentheses):

- (2) word forms left and right from target
- (2) POS sequences left and right bigram around target
- (3) POS tags of left and right word from target, and POS tag of target
- (4) two left and two right nouns from target
- (2) left and right verbs from target
- (2) left and right adjectives from target.

Note that apart from POS tagging, the baseline system does not rely on linguistic preprocessing of the data, and this dependency could have been removed by using the unsupervised POS tagger from Chapter 6. This was not done here, in order to single out the contribution of the WSI features as compared to other systems.

The classification algorithm used in the WSD system is the AODE [245] classifier as provided by the WEKA Machine Learning software [121]. This algorithm is similar to a Naïve Bayes classifier. As opposed to the latter, AODE does not assume mutual independence of features but models correlations between them explicitly, which is desirable in this setting since both baseline and WSI cluster features are expected to be highly inter-correlated. Further, AODE handles symbolic features, so it is possible to use lexical features and cluster IDs directly in the classifier. AODE showed superior performance to other classifiers handling symbolic features in preliminary experiments. For the experiments reported below, 10-fold cross-validation on the training is used for feature selection. Results are reported using held-out test data.

7.4.2 SemEval-07 Lexical Sample Task

The SemEval 2007 lexical sample task (part of task 17) provides 22,281 training sentences for 100 target words, of which 88 (35 nouns and 53 verbs) have more than one sense in the training data. The source for the lexical sample task is the Wall Street Journal, and since the 76,400 sentences from the WSJ Penn Treebank are rather small for a reliable co-occurrence analysis, a 20 million sentence New York Times corpus was used instead.

The sense inventory has been provided by the OntoNotes project [129], using high inter-annotator agreement as a guide for granularity: if fine-grained WordNet senses did not yield high enough agreement with human annotators, they were grouped together. The average number of senses per ambiguous word in the training is 3.95.

10-fold cross validation on the training sentences for ambiguous words was performed, adding all 45 WSI cluster features one at the time. All systems with single cluster features outperformed the baseline system precision of 87.1%, ranging from 88.0%-88.3% precision. For combining the best k single cluster features for $k = 2, 3, 5, 10$, the best performing system resulted in a 10-fold precision of 88.5% for k=3, showing significant gain over the baseline. This system configuration was used for training a system on the full training set. For evaluation, it was applied it to the test data provided by the task organisers.

Since the AODE classifier reports a confidence score (corresponding to the class probability for the winning class at classification time), it is possible to investigate a tradeoff between Precision P and Recall R to optimise the F1-value used for scoring in the lexical sample task. Table 7.4.2 shows the results for the baseline and the system augmented with the top 3 cluster features in comparison with the two best systems in the 2007 evaluation, both for maximal recall and for the optimal F1-value on the test data of 4,851 labelled contexts, which was merely used for evaluation to create the same conditions that held for the participating systems of this task.

It is surprising that the baseline system outperforms the second-best system in the 2007 evaluation. This might be attributed to the AODE classifier used, but also hints at the power of symbolic lexical features in general. The WSI cluster system outperforms the baseline, but does not reach the performance of the winning NUS-ML system. This system uses a variety of standard features plus (global) topic features via LDA and features from dependency parses. However, all reported systems fall into each other's error margins, unlike when evaluating on training data splits. Thus, the WSD setup is competitive to other WSD systems in the literature, while using only minimal linguistic preprocessing and no word sense inventory information beyond what is provided by training examples.

System	NUS-ML	Top3 cluster optimal F1	Top3 cluster max recall	Baseline optimal F1	Baseline max recall	UBC-ALM
F1 value	88.7% $\pm$1.2	88.0% $\pm$1.2	87.8% $\pm$ 1.2	87.5% $\pm$1.2	87.3% $\pm$1.2	86.9% $\pm$1.2

Table 7.1 WSI cluster features and baseline in comparison to the best two systems in the SemEval 2007 Task 17 Lexical Sample evaluation [195]: NUS-ML[51] and UBC-ALM [4]. Error margins provided by the task organisers.

7.4.3 Lexical Substitution System

In this section, a system for lexical substitutions n context is described and evaluated. The system operates on data from the Turk bootstrap Word Sense Inventory (TWSI[2], [30]). This resource consists of over 50,000 sentences. Each sentence contains one out of 397 target nouns annotated by sense. Sense distinctions are drawn according to substitutional equivalence, and substitutions per sense are provided along with the sense definition. The TWSI was created using a bootstrapping process involving human input via three different crowdsourcing tasks. The average number of senses per ambiguous word is 3.91.

Having substitutions associated with word senses in the TWSI, the obvious setup for a lexical substitution system is first disambiguating a target word to assign a TWSI sense, then supplying the associated highest ranked substitutions in context. Evaluation results for both steps are reported separately.

Since TWSI has been created from Wikipedia, an English Wikipedia dump (60 million sentences) from January 2008 is used for gathering co-occurrences for word sense induction.

[2] data available at http://aclweb.org/aclwiki/index.php?title=
TWSI_Turk_bootstrap_Word_Sense_Inventory_%28Repository%29 [August 2011].

7.4.3.1 Word Sense Disambiguation on TWSI data

Figure 8 shows the learning curve for various amounts of randomly selected training/test splits, averaged over three runs each with different random seeds. The most frequent sense (MFS) baseline hovers around 71.5%, the baseline system reaches 80.7% precision at 90% training. For both corpora, the top performing single cluster feature significantly improves over the baseline: The best single cluster feature achieves 82.9% precision. Increasing the number of cluster features leads first to improvements, then to a slight degradation for k=10. The best system using WSI cluster features is obtained by adding the top 5 single WSI cluster features to the baseline features with a 10-fold cross validation score of 83.0%. This system is used in the substitution evaluation experiment described below. While the difference in the 90% training situation is very small, systems using several WSI cluster features excel in reduced-training situations. Looking at the parameterisations of the features, no general trend for either parameter n, t or node weighting was found. Overall, the cluster features reduce the amount of training data needed to reach equal levels of performance to less than half, see Figure 7.3.

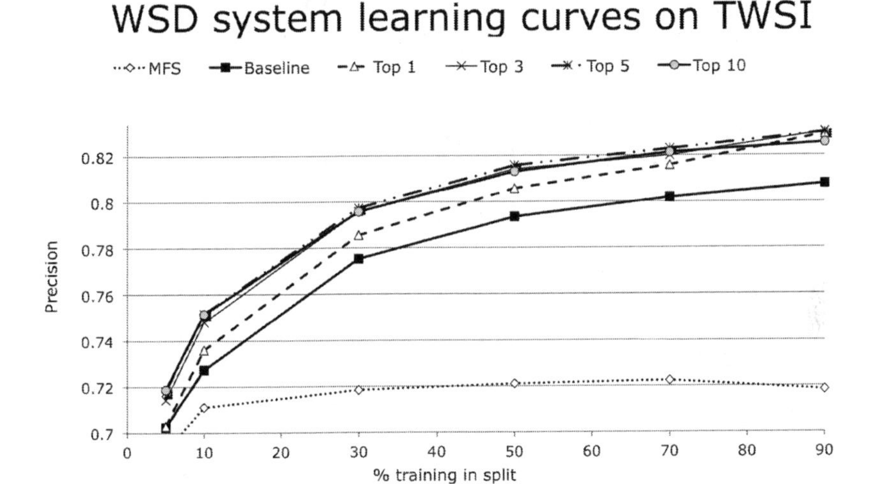

Fig. 7.3 Learning curve on WSD quality for different systems for the TWSI data for 204 ambiguous nouns

7.4.4 *Substitution Acceptability Evaluation*

Now, substitution acceptability of the overall lexical substitution system learnt from TWSI data is evaluated. 500 labelled sentences from the overall data (for all 397 nouns, not just the ambiguous nouns used in the experiments above) were randomly selected. The 10-fold test classifications as described in the previous section were used for word sense assignment. For evaluation, a crowdsourcing task was set up on Amazon Mechanical Turk[3] in the following way: the sentence with the target word in bold is presented along with a) the substitutions for the gold standard sense, b) the substitutions assigned by the system and c) random substitutions from the set of substitutions in separate tasks. Annotators were asked whether the substitutions matched the target word, matched the target word somewhat, do not match the target or the task was impossible for some reason.

Checking for the acceptability of the gold standard substitutions has the following objective: The TWSI sentences with sense labels are not the same sentences for which substitutions were collected [see 30], but rather projected via another crowdsourcing task. The figures reported for a) reflect an upper bound for system-assigned substitutions.

Table 7.2 Substitution acceptability as measured by crowdsourcing for system assignments and random assignments

Answer	a) Gold	b) System	c) Random
YES	469 (93.8%)	456 (91.2%)	12 (2.4%)
NO	14 (2.8%)	27 (5.4%)	485 (97.0%)
SOMEWHAT	17 (3.4%)	17 (3.4%)	3 (0.6%)

Table 7.2 shows the results for averaging over the worker's responses. For being counted as belonging to the YES or NO class, the majority of workers had to choose this option; otherwise the item was counted into the SOMEWHAT class. The gold standard itself is affected by some noise, containing 2.8% errors and 3.4% questionable cases. Despite this, the system is able to assign acceptable substitutions in over 91% of cases, questionable substitutions for 3.4% at an error rate of only 5.4%. Checking the positively judged random assignments, an acceptable substitution was found in about half of the cases by manual inspection, which allows to estimate the worker noise at about 1%. Note that the fraction of acceptable substitutions is only slightly lower than the substitution projection of the TWSI as measured in Section 3.4.

Next, the precision-recall tradeoff in reduced-recall settings is investigated. In applications, it might make sense to only return substitutions when the system is relatively confident. The confidence values of the AODE classifier are used to control recall and provide precision scores for different coverage (percentage of attempted classifications) levels in Table 7.3.

Dependent on the application needs, the confidence score of the classifier allows the system to reduce the error rate of the supplied substitutions. For example at a coverage level of 70%, not even 1% of system substitutions were judged as incorrect.

[3] http://www.mturk.com [August 2011]

Table 7.3 Substitution acceptability in reduced coverage settings. SOMEWHAT class accounts for percentage points missing to 100%.

Coverage	100%	95%	90%	80%	70%
YES	91.2%	91.8%	93.8%	94.8%	95.7%
NO	5.4%	3.4%	2.9%	2.0%	0.9%

7.5 Conclusion on Word Sense Induction and Disambiguation

On the task of semantic disambiguation, it has been demonstrated that yet another time, that Structure Discovery processes capture regularities in large text data in a way that is beneficial for NLP tasks. Again, the contribution of Structure Discovery annotations is not wiped out by, but rather complements standard features for supervised NLP tasks.

Further, with error rates in the single figures and the possibility to reduce errors further by sacrificing recall, the lexical substitution system sketched here constitutes a firm enabling technology for semantic search or other applications.

Chapter 8
Conclusion

Abstract Unsupervised and knowledge-free natural language processing in the Structure Discovery paradigm has shown to be successful and capable of producing a pre-processing quality equal to traditional systems, if just sufficient in-domain raw text can be provided. It is therefore not only a viable alternative for languages with scarce annotated resources, but might also overcome the acquisition bottleneck of language processing for new tasks and applications. In this chapter, the contributions of this book are summarised and put in to a larger perspective. An outlook is given on how Structure Discovery might change the way we design NLP systems in the future.

8.1 Current State of Structure Discovery

This book introduced the Structure Discovery paradigm, which finds and annotates structure in natural language data using Structure Discovery processes. Structure is to be understood here as any kind of annotation that can be automatically introduced into language data by applying algorithms that employ regularities and make these explicit. The algorithms operate without any linguistic information — neither explicitly stated facts about language, nor implicitly encoded knowledge in the form of manual annotations.

The Structure Discovery paradigm was contrasted to the predominant schools of language processing and its adequacy for linguistic phenomena was motivated. A major contribution of this work is the definition of a framework for processing language in an unsupervised and knowledge-free way.

Although there has been previous work that can be attributed to this paradigm, there has not been an attempt to describe it before in its entirety. The main advantages of telling the machine how to discover structure in language is the language- and domain-independence of this approach, which strives at a uniform treatment by employing language universals and reduces the manual effort of encoding language-specific knowledge to zero.

Note that features derived from global statistics, such as e.g. the use of N-grams as features [e.g. 19], is related to Structure Discovery in as much as it employs statistics over a large unannotated text corpus, which informs models for local decisions. But Structure Discovery goes one step further by assigning explicit labels when structural regularities are detected.

When working in Structure Discovery, the first step is to characterise the data that is subject to exploration. At this point, the characteristics are also depending on the representation of the data. Here, graph representations allow to model statistically gathered information about language units in an intuitive way: entities of language are represented by vertices; edge weights denote the degree of association between entities.

It has been known before that many graphs built on language as well as on other data exhibit the scale-free Small World property. After reviewing random graph models and finding that none of the existing models are able to reproduce the characteristics found in word co-occurrence networks, an emergent random text model was presented that comes closer to what is observed in natural language texts. This is reached by simulating the generation of words and sentences in two different modules. Further, the rank-frequency distribution of words, as well as length distributions of words and sentences could be reproduced. This random text generation model is a another step towards explaining the mechanisms that point to the origin of language by defining simple procedures that produce an output stream that quantitatively resembles language.

Clustering methods are the only choice to perform the abstractions and generalisations needed for assigning structure to previously unstructured text in absence of pre-classified items. A review of clustering methods in general and graph clustering methods in particular showed that most previous approaches are not suitable for natural language data, as they can either not cope with highly skewed distributions or fail to present tractable solutions for large datasets.

Thus, the Chinese Whispers graph clustering algorithm was developed and tested on artificial and real datasets. By working in a decentralised fashion and only with local information, this randomised algorithm operates at the lower bound of computational complexity for graph clustering. This allows the processing of very large graphs, which are common in language processing when e.g. encoding words as vertices and their similarities as edges. Several possible extensions and modifications allow adapting the method to special requirements of datasets. A central advantage of this method is that the number of clusters is found automatically rather than provided by the user, which is especially important for language data, where e.g. the number of word senses or the number of languages in a random web sample are not known a priori.

In the practical part of this work, three Structure Discovery processes were set up and thoroughly evaluated.

A language separation algorithm performs almost perfectly in sorting a multilingual text corpus into monolingual chunks. The number and the size distribution of the involved languages are found by the method itself, which renders prior knowledge obsolete for this task.

An unsupervised part-of-speech tagger, which induces and provides annotation with syntactic-semantic word classes, was tested extensively for a variety of languages against manually annotated resources and in applications, where the performance gains are similar to using a traditional tagger. This makes the manual annotation of parts-of-speech information superfluous for application-based settings.

A further process for word sense induction and disambiguation was described, and employed to improve a supervised WSD system as well as to build a high precision lexical substitution system. The training material for the latter was acquired in a data-driven way with crowdsourcing, a further way to reduce the acquisition bottleneck on the side of data-driven task specification.

The feasibility of the broad-scale program of this work — the fully unsupervised and knowledge-free processing of natural language text — has been exemplified on various language processing tasks. It could be demonstrated that in fact Structural Discovery processes can produce structural information that is useful for natural language processing, and are a veritable alternative whenever high manual efforts are not affordable. The presented methods and algorithms serve only as a starting point — further Structure Discovery processes are subject to future work.

The benefit of Structure Discovery for natural language processing is twofold: While Structure Discovery provides a cheap and robust way to rapidly improve the processing for resource-scarce languages and domains, it also allows insights in the mechanisms of natural language per se. For this, the creation of random text models that exhibit the same characteristics as real language — unveiled by SD processes — can be a tool to shed light on the mechanics of language generation and evolution.

The principles of Structure Discovery do not only apply to natural language technology, but also to other naturally occurring phenomena that are not yet understood in its entirety. The study of other aspects of language already benefits from graph-based techniques that adhere to these principles, e.g. for language evolution [61] or for properties of consonant inventories [181].

Algorithms similar in spirit to the ones described here are applied in computational genomics [71]. While its alphabet only consists of four letters, the human genome is — just like language – encoded in linear form and can be interpreted on several levels, like surface realisation, amino acid sequence, or molecule structure. As small-world scale-free graphs are also found in protein interaction, cellular or metabolic networks [233], it can be assumed that similar mechanisms to the ones used for language data can extract useful structural regularities for the understanding of life.

8.2 The Future of Structure Discovery

Where do we go from here? There are several possible directions departing from this point. One direction is clearly the tuning of the SD setup to specific tasks, by systematically overgenerating SD annotations stemming from different methodologies and different parameterisations. This can be viewed as supervision, since SD methods are evaluated for their suitability for given tasks — nevertheless, this supervision is indirect since it is not a specific feature set that is optimised, but the parameterisation of the method used to produce this feature set. It should be stressed again that Structure Discovery is not at all restricted to the graph representation used in this book, and Structure Discovery encourages the use of diverse methods and representations to harvest their respective advantages in a common framework. Promising approaches in this respect are topic models [133], approaches to unsupervised parsing [36, inter al.], creative uses of distributional similarity [153] and methods to automatically structure domain knowledge by identifying templates [56], just to name a few.

A very important next step is to move from words to larger units. This starts with the identification of multiword expressions (as in groups of words that have a certain binding strength). Since the definition of what constitutes a multiword is application-specific, this should entail either different kinds of multiwords, a graded score of multiwordness [cf. 31], or a combination of both. Using similarity measures on phrases gives rise to Structure Discovery for paraphrase detection, which in turn could improve document classification and summarisation, as demonstrated in the graph-based systems of Mihalcea and Tarau [175] and Erkan and Radev [90].

But Structure Discovery does not stop at aggregating words or multiwords into equivalence classes: also, relations between them can be detected, as Turney [238] has demonstrated. This gives rise to the learning of lightweight ontology-like structures, which again add layers of structure, as well as enable the extraction of 'typed' facts and instances, without specifying the target types in advance. While the success of IBM's Watson system in the Jeopardy! competition was impressive, wouldn't it even be more impressive if most or all of the heterogenous parts of the system were automatically induced, and merely tuned to the task by correct question-answer pairs?

This opens up an entirely new way of constructing systems that perform language processing tasks. The current approach of concatenating standard preprocessing steps in a pipeline is replaced by merely providing data of two kinds. Task data defines the task being solved, for example question answer pairs, relevant documents for queries, or multi-document summaries. Task data does not need to be static, but can also be acquired during an active learning process or via user interaction. The other kind of data, large unannotated raw text, implicitly provides everything from language structure to domain knowledge. From a plethora of Structure Discovery processes that are stored in a repository and that are instantiated by running them over this raw data, those processes are selected that most benefit a system that solves the task. At this, not the single parts of a pipeline are optimised, but rather their interplay. Thus, this very generic system becomes reactive to the task defined by the task data and learns to solve tasks without the need to manually tune its inner workings. Surely, it will be hard to understand such a system in detail. But who would want to meticulously control every piece of such a system, when one can simply let it emerge?

References

[1] Abney, S.: Statistical Methods and Linguistics. In: Klavans, J., Resnik, P. (eds.) The Balancing Act: Combining Symbolic and Statistical Approaches to Language, pp. 1–26. The MIT Press, Cambridge, Massachusetts (1996)

[2] Adamic, L.A.: Zipf, Power-law, Pareto – a ranking tutorial. Tech. rep., Information Dynamics Lab, HP Labs, HP Labs, Palo Alto, CA 94304 (October 2000)

[3] Agirre, E., Edmonds, P. (eds.): Word Sense Disambiguation: Algorithms and Applications. Text, Speech and Language Technology, Springer (July 2006)

[4] Agirre, E., de Lacalle, O.L.: UBC-ALM: combining k-NN with SVD for WSD. In: Proceedings of the 4th International Workshop on Semantic Evaluations. pp. 342–345. SemEval '07, Association for Computational Linguistics, Stroudsburg, PA, USA (2007)

[5] Agirre, E., Màrquez, L., Wicentowski, R. (eds.): Proceedings of the Fourth International Workshop on Semantic Evaluations (SemEval-2007). Association for Computational Linguistics, Prague, Czech Republic (June 2007)

[6] Agirre, E., Martínez, D., López de Lacalle, O., Soroa, A.: Evaluating and optimizing the parameters of an unsupervised graph-based WSD algorithm. In: Proceedings of TextGraphs: the Second Workshop on Graph Based Methods for Natural Language Processing. pp. 89–96. Association for Computational Linguistics, New York City (June 2006)

[7] Agirre, E., Martínez, D., López de Lacalle, O., Soroa, A.: Two graph-based algorithms for state-of-the-art WSD. In: Proceedings of the Conference on Empirical Methods in Natural Language Processing (EMNLP-06). pp. 585–593. Association for Computational Linguistics, Sydney, Australia (July 2006)

[8] Aiello, W., Chung, F., Lu, L.: A random graph model for massive graphs. In: STOC '00: Proceedings of the thirty-second annual ACM symposium on Theory of computing. pp. 171–180. ACM Press, New York, NY, USA (2000)

[9] Albert, R., Jeong, H., Barabási, A.L.: Error and attack tolerance of complex networks. Nature 406, 378 (2000)

[10] Allauzen, C., Mohri, M., Roark, B.: Generalized algorithms for constructing statistical language models. In: Proceedings of the 41st Annual Meeting on Association for Computational Linguistics (ACL-03). pp. 40–47. Association for Computational Linguistics, Morristown, NJ, USA (2003)
[11] Amaral, L.A., Scala, A., Barthelemy, M., Stanley, H.E.: Classes of small-world networks. Proc. Natl. Acad. Sci. USA 97(21) (2000)
[12] Amblard, F.: Which ties to choose? A survey of social networks models for agent-based social simulations. In: Proceedings of the 2002 SCS International Conference On Artificial Intelligence, Simulation and Planning in High Autonomy Systems. pp. 253–258 (2002)
[13] Baeza-Yates, R., Ribeiro-Neto, B.: Modern Information Retrieval. The Concepts and Technology behind Search (2nd Edition). ACM Press Books (2009)
[14] Baker, C., Fillmore, C., Cronin, B.: The structure of the FrameNet database. International Journal of Lexicography 16(3), 281–296 (2003)
[15] Banko, M., Brill, E.: Scaling to Very Very Large Corpora for Natural Language Disambiguation. In: Proceedings of ACL-01. pp. 26–33 (2001)
[16] Barabási, A.L., Albert, R.: Emergence of scaling in random networks. Science 286, 509 (1999)
[17] Barrat, A., Barthelemy, M., Pastor-Satorras, R., Vespignani, A.: The architecture of complex weighted networks. Proc. Natl. Acad. Sci. USA 101, 37–47 (2004)
[18] Barrat, A., Barthelemy, M., Vespignani, A.: Weighted evolving networks: coupling topology and weights dynamics. Physical Review Letters 92 (2004)
[19] Bergsma, S., Pitler, E., Lin, D.: Creating robust supervised classifiers via web-scale n-gram data. In: Proceedings of the 48th Annual Meeting of the Association for Computational Linguistics. pp. 865–874. ACL '10, Association for Computational Linguistics, Stroudsburg, PA, USA (2010)
[20] Berkhin, P.: Survey Of Clustering Data Mining Techniques. Tech. rep., Accrue Software, San Jose, CA (2002)
[21] Berland, M., Charniak, E.: Finding parts in very large corpora. In: Proceedings of the 37th annual meeting of the Association for Computational Linguistics (ACL-99). pp. 57–64. Association for Computational Linguistics, Morristown, NJ, USA (1999)
[22] Bern, M., Eppstein, D.: Approximation algorithms for geometric problems. In: Approximation algorithms for NP-hard problems, pp. 296–345. PWS Publishing Co., Boston, MA, USA (1997)
[23] Biemann, C.: Bootstrapping. In: Heyer, G., Quasthoff, U., Wittig, T. (eds.) Wissensrohstoff Text, pp. 260–266. W3L, Bochum (2006)
[24] Biemann, C., Böhm, C., Heyer, G., Melz, R.: Automatically Building Concept Structures and Displaying Concept Trails for the Use in Brainstorming Sessions and Content Management Systems. In: Proceedings of Innovative Internet Community Systems (IICS-2004). Springer LNCS, Guadalajara, Mexico (2004)
[25] Biemann, C., Bordag, S., Quasthoff, U.: Automatic Acquisition of Paradigmatic Relations using Iterated Co-occurrences. In: Proceedings of the fourth

international conference on Language Resources and Evaluation (LREC-04). Lisbon, Portugal (2004)

[26] Biemann, C., Quasthoff, U.: Similarity of Documents and Document Collections using Attributes with Low Noise. In: Proceedings of the Third International Conference on Web Information Systems and Technologies (WEBIST-07). pp. 130–135. Barcelona, Spain (2007)

[27] Biemann, C.: Chinese Whispers – an Efficient Graph Clustering Algorithm and its Application to Natural Language Processing Problems. In: Proceedings of TextGraphs: the Second Workshop on Graph Based Methods for Natural Language Processing. pp. 73–80. Association for Computational Linguistics, New York City (June 2006)

[28] Biemann, C.: A Random Text Model for the Generation of Statistical Language Invariants. In: Human Language Technologies 2007: The Conference of the North American Chapter of the Association for Computational Linguistics; Proceedings of the Main Conference (HLT-NAACL-07). pp. 105–112. Association for Computational Linguistics, Rochester, New York (April 2007)

[29] Biemann, C.: Unsupervised Part-of-Speech Tagging in the Large. Research on Language and Computation 7, 101–135 (2009)

[30] Biemann, C.: Creating a System for Lexical Substitutions from Scratch using Crowdsourcing. Language Resources and Evaluation: Special Issue on Collaboratively Constructed Language Resources (to appear 2011)

[31] Biemann, C., Giesbrecht, E.: Distributional semantics and compositionality 2011: Shared task description and results. In: Proceedings of the Workshop on Distributional Semantics and Compositionality. pp. 21–28. Association for Computational Linguistics, Portland, Oregon, USA (June 2011)

[32] Biemann, C., Giuliano, C., Gliozzo, A.: Unsupervised Part-of-Speech Tagging Supporting Supervised Methods. In: Proceedings of Recent Advances in Natural Language Processing (RANLP-07). Borovets, Bulgaria (2007)

[33] Biemann, C., Shin, S.I., Choi, K.S.: Semiautomatic extension of CoreNet using a bootstrapping mechanism on corpus-based co-occurrences. In: Proceedings of the 20th international conference on Computational Linguistics (COLING-04). Association for Computational Linguistics, Morristown, NJ, USA (2004)

[34] Biemann, C., Teresniak, S.: Disentangling from Babylonian Confusion – Unsupervised Language Identification. In: Proceedings of Computational Linguistics and Intelligent Text Processing, 6th International Conference (CICLing-05). pp. 773–784. Springer LNCS, Mexico D.F., Mexico (2005)

[35] Blei, D.M., Ng, A.Y., Jordan, M.I.: Latent dirichlet allocation. Journal of Machine Learning Research 3, 993–1022 (2003)

[36] Bod, R.: An all-subtrees approach to unsupervised parsing. In: Proceedings of the 21st International Conference on Computational Linguistics and the 44th annual meeting of the ACL (COLING/ACL-06). pp. 865–872. Association for Computational Linguistics, Morristown, NJ, USA (2006)

[37] Bollobas, B., Riordan, O., Spencer, J., Tusnady, G.: The degree sequence of a scale-free random graph process. Random Structures and Algorithms 18(3) (2001)
[38] Bollobas, B.: Random Graphs. Academic Press (1985)
[39] Bollobas, B.: Modern Graph Theory. Springer-Verlag (July 1998)
[40] Bonato, A.: A Survey of Models of the Web Graph. Combinatorial and Algorithmic Aspects of Networking pp. 159–172 (2005)
[41] Bordag, S.: Word sense induction: Triplet-based clustering and automatic evaluation. In: Proceedings of the 11th Conference of the European Chapter of the Association for Computational Linguistics (EACL-06). Trento, Italy (2006)
[42] Bordag, S.: Elements of Knowledge-free and Unsupervised Lexical Acquisition. Ph.D. thesis, University of Leipzig (2007)
[43] van den Bosch, A., Buchholz, S.: Shallow parsing on the basis of words only: a case study. In: ACL '02: Proceedings of the 40th Annual Meeting on Association for Computational Linguistics. pp. 433–440. Association for Computational Linguistics, Morristown, NJ, USA (2001)
[44] Brandes, U., Gaertler, M., Wagner, D.: Experiments on graph clustering algorithms. In: Proceedings of the 11th Annual European Symposium on Algorithms (ESA'03). pp. 568–579. Springer (2003)
[45] Brants, T.: TnT: a statistical part-of-speech tagger. In: Proceedings of the sixth conference on Applied Natural Language Processing (ANLP-00)). pp. 224–231. Morgan Kaufmann Publishers Inc., San Francisco, CA, USA (2000)
[46] Brants, T., Hendriks, R., Kramp, S., Krenn, B., Preis, C., Skut, W., Uszkoreit, H.: Das NEGRA-Annotationsschema. Negra project report, Universität des Saarlandes, Saarbrücken (1997), http://www.coli.uni-sb.de/sfb378/negra-corpus/negra-corpus.html
[47] Brill, E.: A simple rule-based part of speech tagger. In: Proceedings of the third conference on Applied Natural Language Processing (ANLP-92)). pp. 152–155. Association for Computational Linguistics, Morristown, NJ, USA (1992)
[48] Brin, S., Page, L.: The anatomy of a large-scale hypertextual Web search engine. Computer Networks and ISDN Systems 30(1-7), 107–117 (1998)
[49] Brown, P.F., Pietra, V.J.D., deSouza, P.V., Lai, J.C., Mercer, R.L.: Class-Based n-gram Models of Natural Language. Computational Linguistics 18(4), 467–479 (1992)
[50] Burnard, L.: Users Reference Guide for the British National Corpus. Oxford University Computing Service, Oxford, U.K. (1995)
[51] Cai, J.F., Lee, W.S., Teh, Y.W.: NUS-ML: Improving word sense disambiguation using topic features. In: Proceedings of the 4th International Workshop on Semantic Evaluations. pp. 524–531. SemEval '07, Association for Computational Linguistics (2007)
[52] Callison-Burch, C., Dredze, M.: Creating Speech and Language Data With Amazon's Mechanical Turk. In: Proceedings of the NAACL HLT 2010 Work-

shop on Creating Speech and Language Data with Amazon's Mechanical Turk. pp. 1–12. Los Angeles, CA, USA (2010)
[53] Cardie, C., Weischedel, R. (eds.): A Corpus-Based Approach for Building Semantic Lexicons. Association for Computational Linguistics, Somerset, New Jersey (1997)
[54] Cavnar, W.B., Trenkle, J.M.: N-Gram-Based Text Categorization. In: Proceedings of SDAIR-94, 3rd Annual Symposium on Document Analysis and Information Retrieval. pp. 161–175. Las Vegas, US (1994)
[55] Chakrabarti, D., Faloutsos, C.: Graph mining: Laws, generators, and algorithms. ACM Comput. Surv. 38(1) (2006)
[56] Chambers, N., Jurafsky, D.: Template-Based Information Extraction without the Templates. In: Proceedings of the 49th Annual Meeting of the Association for Computational Linguistics: Human Language Technologies. pp. 976–986. Association for Computational Linguistics, Portland, Oregon, USA (June 2011)
[57] Charniak, E., Hendrickson, C., Jacobson, N., Perkowitz, M.: Equations for Part-of-Speech Tagging. In: National Conference on Artificial Intelligence. pp. 784–789 (1993)
[58] Chomsky, N.: Syntactic Structures. Mouton, The Hague (1957)
[59] Chomsky, N.: Current Issues in Linguistic Theory. Mouton, The Hague (1964)
[60] Choudhury, M., Chatterjee, D., Mukherjee, A.: Global topology of word co-occurrence networks: Beyond the two-regime power-law. In: Coling 2010: Posters. pp. 162–170. Coling 2010 Organizing Committee, Beijing, China (August 2010)
[61] Choudhury, M., Jalan, V., Sarkar, S., Basu, A.: Evolution, optimization, and language change: the case of bengali verb inflections. In: Proceedings of Ninth Meeting of the ACL Special Interest Group in Computational Morphology and Phonology. pp. 65–74. SigMorPhon '07, Association for Computational Linguistics, Stroudsburg, PA, USA (2007)
[62] Christiano, P., Kelner, J.A., Madry, A., Spielman, D.A., Teng, S.H.: Electrical flows, Laplacian systems, and faster approximation of maximum flow in undirected graphs. In: Proceedings of the 43rd ACM Symposium on Theory of Computing, STOC 2011. pp. 273–282. San Jose, CA, USA (June 2011)
[63] Chung, F.R.: Spectral Graph Theory. Regional Conference Series in Mathematics 92 (1997)
[64] Church, K.W.: Empirical estimates of adaptation: the chance of two noriegas is closer to p/2 than p^2. In: Proceedings of the 18th conference on Computational Linguistics (COLING-00). pp. 180–186. Association for Computational Linguistics, Morristown, NJ, USA (2000)
[65] Clark, A.: Inducing Syntactic Categories by Context Distribution Clustering. In: Cardie, C., Daelemans, W., Nédellec, C., Sang, E.T.K. (eds.) Proceedings of the Fourth Conference on Computational Natural Language Learning and of the Second Learning Language in Logic Workshop, Lisbon, 2000, pp. 91–94. Association for Computational Linguistics, Somerset, New Jersey (2000)

[66] Clark, A.: Combining distributional and morphological information for part of speech induction. In: Proceedings of the tenth conference on European chapter of the Association for Computational Linguistics (EACL-03). pp. 59–66. Association for Computational Linguistics, Morristown, NJ, USA (2003)

[67] Cohen, P.R., Adams, N., Heeringa, B.: Voting Experts: An Unsupervised Algorithm for Segmenting Sequences. International Journal on Intelligent Data Analysis 11 (2007)

[68] Cohn, H., Kleinberg, R., Szegedy, B., Umans, C.: Group-theoretic Algorithms for Matrix Multiplication. In: FOCS '05: Proceedings of the 46th Annual IEEE Symposium on Foundations of Computer Science. pp. 379–388. IEEE Computer Society, Washington, DC, USA (2005)

[69] Cohn, T.: Performance Metrics for Word Sense Disambiguation. In: Proceedings of the Australasian Language Technology Workshop, 2003. pp. 49–56 (2003)

[70] da Costa, L., Rodrigues, F.A., Travieso, G., Boas, P.R.V.: Characterization of complex networks: A survey of measurements. Advances in Physics 56(1), 167–242 (January 2005)

[71] Cristianini, N., Hahn, M.W.: Introduction to Computational Genomics: A Case Studies Approach. Cambridge University Press (2007)

[72] Cruse, D.A.: Aspects of the Micro-Structure of Word Meanings. In: Ravin, Y., Leacock, C. (eds.) Polysemy: Theoretical and Computational Approaches. Oxford. University Press (2002)

[73] Cucerzan, S., Yarowsky, D.: Language independent named entity recognition combining morphological and contextual evidence. In: Proceedings of 1999 Joint SIGDAT Conference on EMNLP and VLC . pp. 132–138. College Park (1999)

[74] Cvetkovič, D., Doob, M., Sachs, H.: Spectra of Graphs: Theory and Application (3rd Edition). Johann Ambrosius Barth (1995)

[75] Cysouw, M., Biemann, C., Ongyerth, M.: Using Strong's Numbers in the Bible to Test an Automatic Alignment of Parallel Texts. Special issue of Sprachtypologie und Universalienforschung (STUF) pp. 66–79 (2007)

[76] Daumé, H., Marcu, D.: Domain Adaptation for Statistical Classifiers. J. Artif. Intell. Res. (JAIR) 26, 101–126 (2006)

[77] Davidov, D., Rappoport, A.: Efficient Unsupervised Discovery of Word Categories Using Symmetric Patterns and High Frequency Words. In: Proceedings of the 21st International Conference on Computational Linguistics and 44th Annual Meeting of the Association for Computational Linguistics (COLING/ACL-06). pp. 297–304. Association for Computational Linguistics, Sydney, Australia (July 2006)

[78] Davidov, D., Rappoport, A., Koppel, M.: Fully Unsupervised Discovery of Concept-Specific Relationships by Web Mining. In: Proceedings of the 45th Annual Meeting of the Association of Computational Linguistics (ACL-07). pp. 232–239. Association for Computational Linguistics, Prague, Czech Republic (June 2007)

[79] Deerwester, S., Dumais, S.T., Furnas, G.W., Landauer, T.K., Harshman, R.: Indexing by Latent Semantic Analysis. Journal of the American Society for Information Science 41(6), 391–407 (January 1990)
[80] Deo, N., Gupta, P.: World Wide Web: a Graph-Theoretic perspective. Tech. rep., Comp. Sci. Tech. Report CS-TR-01-001, University of Central Florida (2001)
[81] Dhillon, I.S., Mallela, S., Modha, D.S.: Information-Theoretic Co-Clustering. In: Proceedings of The Ninth ACM SIGKDD International Conference on Knowledge Discovery and Data Mining(KDD-2003). pp. 89–98 (2003)
[82] van Dongen, S.M.: Graph Clustering by Flow Simulation. Ph.D. thesis, University of Utrecht, Netherlands (2000)
[83] Dorogovtsev, S.N., Mendes, J.F.F.: Language as an evolving word web. Proceedings of The Royal Society of London. Series B, Biological Sciences 268(1485), 2603–2606 (December 2001)
[84] Dorogovtsev, S.N., Mendes, J.F.F.: Scaling properties of scale-free evolving networks: Continuous approach. Physical Review E 63 (2001)
[85] Dunning, T.: Statistical identification of language. Techical Report MCCS-94-273, Computing Research Lab (CRL), New Mexico State University (1994)
[86] Dunning, T.E.: Accurate Methods for the Statistics of Surprise and Coincidence. Computational Linguistics 19(1), 61–74 (1993)
[87] Eiken, U.C., Liseth, A.T., Witschel, H.F., Richter, M., Biemann, C.: Ord i dag: Mining Norwegian Daily Newswire. In: Proceedings of the FinTAL. Turku, Finland (2006)
[88] Erdős, P., Rényi, A.: On the evolution of random graphs. Publ. Math. Inst. Hungar. Acad. Sci. 5, 17–61 (1960)
[89] Erk, K., Strapparava, C. (eds.): Proceedings of the 5th International Workshop on Semantic Evaluation. Association for Computational Linguistics, Uppsala, Sweden (July 2010)
[90] Erkan, G., Radev, D.R.: LexRank: Graph-based Lexical Centrality as Salience in Text Summarization. Journal of Artificial Intelligence Research (JAIR) 22, 457–479 (2004)
[91] Ertöz, L., Steinbach, M., Kumar, V.: A new shared nearest neighbor clustering algorithm and its applications. In: Proceedings of Workshop on Clustering High Dimensional Data and its Applications. pp. 105–115 (2002)
[92] Ester, M., Kriegel, H.P., Sander, J., Xu, X.: A density-based algorithm for discovering clusters in large spatial databases with noise. In: Proceedings of 2nd International Conference on Knowledge Discovery and Data Mining (KDD'96). AAAI press, Portland, Oregon (1996)
[93] Everitt, B.S., Landau, S., Leese, M.: Cluster analysis (4th edition). Arnold (2001)
[94] Evert, S.: The Statistics of Word Co-occurrences: Word Pairs and Collocations. Ph.D. thesis, University of Stuttgart (2004)

[95] Ferrer-i-Cancho, R., Solé, R.V.: The small world of human language. Proceedings of The Royal Society of London. Series B, Biological Sciences 268(1482), 2261–2265 (November 2001)
[96] Ferrer-i-Cancho, R., Solé, R.V.: Zipf's law and random texts. Advances in Complex Systems 5(1), 1–6 (2002)
[97] Ferrucci, D.: Building Watson: An Overview of the DeepQA Project. Invited talk at ACL 2011, Portland, Oregon, USA (2011)
[98] Finch, S., Chater, N.: Bootstrapping syntactic categories using statistical methods. In: Background and Experiments in Machine Learning of Natural Language: Proceedings of the 1st SHOE Workshop. pp. 229–235. Katholieke Universiteit, Brabant, Holland (1992)
[99] Firth, J.R.: A Synopsis of Linguistic Theory, 1933-1955. Blackwell, Oxford (1957)
[100] Fjällström, P.O.: Algorithms for graph partitioning: A survey. Linkoping Electronic Articles in Computer and Information Science 3 (1998)
[101] Flake, G.W., Tarjan, R.E., Tsioutsiouliklis, K.: Graph Clustering and Minimum Cut Trees. Internet Mathematics 1(4), 385–408 (2004)
[102] Ford, L.R., Fulkerson, D.R.: Maximal flow through a network. Canadian Journal of Mathematics 8, 399–404 (1956)
[103] Francis, W.N., Kučera, H.: Frequency Analysis of English Usage. Houghton Mifflin, Boston (1982)
[104] Freitag, D.: Toward unsupervised whole-corpus tagging. In: Proceedings of the 20th international conference on Computational Linguistics (COLING-04). p. 357. Association for Computational Linguistics, Morristown, NJ, USA (2004)
[105] Freitag, D.: Trained named entity recognition using distributional clusters. In: Proceedings of EMNLP-04. pp. 262–269 (2004)
[106] Garside, R., Leech, G., Sampson, G.: The computational analysis of English: a corpus-based approach. Longman (1987)
[107] Gauch, S., Futrelle, R.: Experiments in Automatic Word Class and Word Sense Identification for Information Retrieval. In: Proceedings of the 3rd Annual Symposium on Document Analysis and Information Retrieval. pp. 425–434. Las Vegas, NV (April 1994)
[108] Gibbons, A.: Algorithmic Graph Theory. Cambridge University Press (1985)
[109] Gimpel, K., Schneider, N., O'Connor, B., Das, D., Mills, D., Eisenstein, J., Heilman, M., Yogatama, D., Flanigan, J., Smith, N.A.: Part-of-Speech Tagging for Twitter: Annotation, Features, and Experiments. In: Proceedings of the 49th Annual Meeting of the Association for Computational Linguistics: Human Language Technologies. pp. 42–47. Association for Computational Linguistics, Portland, Oregon, USA (June 2011)
[110] Glassman, S.: A caching relay for the World Wide Web. Computer Networks and ISDN Systems 27(2), 165–173 (1994)
[111] Gliozzo, A.M., Giuliano, C., Strapparava, C.: Domain Kernels for Word Sense Disambiguation. In: Proceedings of the 43rd Annual Meeting of the

Association for Computational Linguistics (ACL-05). pp. 403–410. Ann Arbor, Michigan, USA (2005)
[112] Gliozzo, A.M., Strapparava, C.: Semantic Domains in Computational Linguistics. Springer (2009)
[113] Goldwater, S., Griffiths, T.L., Johnson, M.: Producing power-law distributions and damping word frequencies with two-stage language models. J. Mach. Learn. Res. 999999, 2335–2382 (July 2011), http://dl.acm.org/citation.cfm?id=2021026.2021075
[114] Goldwater, S., Griffiths, T.: A fully Bayesian approach to unsupervised part-of-speech tagging. In: Proceedings of the 45th Annual Meeting of the Association of Computational Linguistics. pp. 744–751. Association for Computational Linguistics, Prague, Czech Republic (June 2007)
[115] Grefenstette, G.: Comparing Two Language Identification Schemes. In: 3rd International Conference on Statistical Analysis of Textual Data. pp. 263–268. Rome, Italy (1995)
[116] Grewendorf, G., Hamm, F., Sternefeld, W.: Sprachliches Wissen. Suhrkamp, Frankfurt/Main (1987)
[117] Ha, L.Q., Sicilia-Garcia, E.I., Ming, J., Smith, F.J.: Extension of Zipf's law to words and phrases. In: Proceedings of the 19th international conference on Computational Linguistics (COLING-02). pp. 1–6. Association for Computational Linguistics, Morristown, NJ, USA (2002)
[118] Hagen, K., Johannessen, J.B., Nøklestad, A.: A Constraint-Based Tagger for Norwegian. Lindberg, C.-E. og S. Nordahl Lund (red.): 17th Scandinavian Conference of Linguistics, vol. I. Odense: Odense Working Papers in Language and Communication I(19) (2000)
[119] Haghighi, A., Klein, D.: Prototype-Driven Learning for Sequence Models. In: Proceedings of the Human Language Technology Conference of the North American Chapter of the Association of Computational Linguistics (HLT-NAACL-06). New York, NY, USA (2006)
[120] Haghighi, A., Klein, D.: Unsupervised Coreference Resolution in a Nonparametric Bayesian Mode. In: Proceedings of the 45th Annual Meeting of the Association of Computational Linguistics (ACL-07). pp. 848–855. Association for Computational Linguistics, Prague, Czech Republic (June 2007)
[121] Hall, M., Frank, E., Holmes, G., Pfahringer, B., Reutemann, P., Witten, I.H.: The WEKA data mining software: an update. SIGKDD Explor. Newsl. 11, 10–18 (November 2009)
[122] Han, B., Baldwin, T.: Lexical Normalisation of Short Text Messages: Makn Sens a #twitter. In: Proceedings of the 49th Annual Meeting of the Association for Computational Linguistics: Human Language Technologies. pp. 368–378. Association for Computational Linguistics, Portland, Oregon, USA (June 2011)
[123] Harris, Z.S.: Mathematical Structures of Language. Wiley (1968)
[124] Harris, Z.S.: Methods in Structural Linguistics. University of Chicago Press, Chicago (1951)

[125] Hauck, S., Borriello, G.: An evaluation of bipartitioning techniques. In: ARVLSI '95: Proceedings of the 16th Conference on Advanced Research in VLSI (ARVLSI'95). p. 383. IEEE Computer Society, Washington, DC, USA (1995)
[126] Heyer, G., Quasthoff, U., Wittig, T.: Wissensrohstoff Text. W3L, Bochum, Bochum, Germany (2006)
[127] Holz, F., Witschel, H., Heinrich, G., Heyer, G., Teresniak, S.: An evaluation framework for semantic search in P2P networks. In: Proceedings of the Seventh International Workshop on Innovative Internet Community Systems, (I2CS07). Munich, Germany (2007)
[128] Horn, G.M.: Lexical-Functional Grammar. Mouton de Gruyter, Berlin (1983)
[129] Hovy, E., Marcus, M., Palmer, M., Ramshaw, L., Weischedel, R.: OntoNotes: The 90% solution. In: Proceedings of HLT-NAACL 2006. pp. 57–60 (2006)
[130] Jäger, G.: Evolutionary Game Theory and Linguistic Typology: A Case Study. In: Dekker, P. (ed.) Proceedings of the 14th Amsterdam Colloquium. ILLC, University of Amsterdam (2003)
[131] Jain, A.K., Murty, M.N., Flynn, P.J.: Data clustering: a review. ACM Computing Surveys 31(3), 264–323 (1999)
[132] Johnson, S.: Solving the problem of language recognition. Tech. rep., School of Computer Studies, University of Leeds, UK (1993)
[133] Jordan, M.I.: Graphical Models. Statist. Sci. 19(1), 140–155 (2004)
[134] Kannan, R., Vempala, S., Vetta, A.: On Clusterings: Good, Bad, and Spectral. In: Proceedings of the 41st Annual Symposium on the Foundation of Computer Science. pp. 367–380. IEEE Computer Society (Nov 2000)
[135] Kanter, I., Kessler, D.A.: Markov Processes: Linguistics and Zipf's Law. Physical Review Letters 74(22), 4559–4562 (May 1995)
[136] Karger, D.R.: Minimum cuts in near-linear time. In: STOC '96: Proceedings of the twenty-eighth annual ACM symposium on Theory of computing. pp. 56–63. ACM Press, New York, NY, USA (1996)
[137] Kay, M.: Machine translation: the disappointing past and present. In: Survey of the state of the art in human language technology, pp. 248–250. Cambridge University Press, New York, NY, USA (1997)
[138] Keller, E.F.: Revisiting "scale-free" networks. Bioessays 27(10), 1060–1068 (October 2005)
[139] Kilgarriff, A.: I Don't Belive in Word Senses. Computers and the Humanities 31(2), 91–113 (1997)
[140] Klein, D.: The Unsupervised Learning of Natural Language Structure. Ph.D. thesis, Stanford University (2005)
[141] Kleinberg, J.M.: Authoritative sources in a hyperlinked environment. Journal of the ACM 46(5), 604–632 (1999)
[142] Kleinberg, J.M., Kumar, R., Raghavan, P., Rajagopalan, S., Tomkins, A.S.: The Web as a Graph: Measurements, Models and Methods. In: Proceedings of the International Conference on Combinatorics and Computing. pp. 1–17. Singapore (1999)

[143] Knuth, D.E.: The Art of Computer Programming, Volume 3: (2nd Ed.) Sorting and Searching. Addison Wesley Longman Publishing Co., Inc., Redwood City, CA, USA (1998)
[144] Kohavi, R., John, G.H.: Wrappers for feature subset selection. Artificial Intelligence 97(1-2), 273–324 (1997)
[145] Kumar, S.R., Raghavan, P., Rajagopalan, S., Tomkins, A.: Extracting Large-Scale Knowledge Bases from the Web. In: The VLDB Journal. pp. 639–650 (1999)
[146] Kurimo, M., Virpioja, S., Turunen, V., Lagus, K.: Morpho Challenge competition 2005–2010: evaluations and results. In: Proceedings of the 11th Meeting of the ACL Special Interest Group on Computational Morphology and Phonology. pp. 87–95. SIGMORPHON '10, Association for Computational Linguistics (2010)
[147] Ladefoged, P., Maddieson, I.: The Sounds of the World's Languages. Blackwell Publishers, Oxford, UK (1996)
[148] Lafferty, J., McCallum, A., Pereira, F.: Conditional Random Fields: Probabilistic Models for Segmenting and Labeling Sequence Data. In: Proceedings of the 18th International Conference on Machine Learning (ICML-01). pp. 282–289. Morgan Kaufmann, San Francisco, CA (2001)
[149] Lempel, R., Moran, S.: Predictive caching and prefetching of query results in search engines. In: Proceedings of the 12th international conference on World Wide Web (WWW-03). pp. 19–28. ACM Press, New York, NY, USA (2003)
[150] Levin, E., Sharifi, M., Ball, J.: Evaluation of Utility of LSA for Word Sense Discrimination. In: Proceedings of the Human Language Technology Conference of the NAACL, Companion Volume: Short Papers. pp. 77–80. Association for Computational Linguistics, New York City, USA (June 2006)
[151] Li, C., Chen, G.: Network connection strengths: Another power-law? Tech. rep., cond-mat/0311333, ArXiv, 2003 (2003)
[152] Li, W.: Random texts exhibit zipf's law-like word frequency distribution. IEEETIT: IEEE Transactions on Information Theory 38 (1992)
[153] Lin, D.: Automatic Retrieval and Clustering of Similar Words. In: Proceedings of the 36th Annual Meeting of the Association for Computational Linguistics and 17th International Conference on Computational Linguistics, Volume 2. pp. 768–774. Association for Computational Linguistics, Montreal, Quebec, Canada (August 1998)
[154] Lu, L., Zhou, T.: Link Prediction in Complex Networks: A Survey. arXiv, http://arxiv.org/abs/1010.0725 (Oct 2010)
[155] MacQueen, J.: Some methods for classification and analysis of multivariate observations. In: Proceedings of the Fifth Berkeley Symposium on Mathematical Statistics and Probability, volume I. pp. 281–297. Berkeley University of California Press (1967)
[156] Mahn, M., Biemann, C.: Tuning Co-occurrences of Higher Orders for Generating Ontology Extension Candidates. In: Proceedings of the ICML-05

Workshop on Ontology Learning and Extension using Machine Learning Methods. Bonn, Germany (2005)
[157] Mandelbrot, B.B.: An information theory of the statistical structure of language. In: Proceedings of the Symposium on Applications of Communications Theory. Butterworths (1953)
[158] Manning, C.D., Schütze, H.: Foundations of Statistical Natural Language Processing. The MIT Press, Cambridge, Massachusetts (1999)
[159] Martínez, D., de Lacalle, O.L., Agirre, E.: On the use of automatically acquired examples for all-nouns word sense disambiguation. J. Artif. Int. Res. 33, 79–107 (September 2008)
[160] Matsumoto, M., Nishimura, T.: Mersenne twister: a 623-dimensionally equidistributed uniform pseudo-random number generator. ACM Trans. Model. Comput. Simul. 8(1), 3–30 (1998)
[161] Matsuo, Y.: Clustering Using Small World Structure. In: Proceedings of the 6th International Conference on Knowledge-based Intelligent Information Engineering Systems and Applied Technologies (KES-02). pp. 1257–1261. IOS Press/Ohmsha, Crema, Italy (2002)
[162] McCallum, A.K.: Mallet: A machine learning for language toolkit (2002), http://mallet.cs.umass.edu
[163] McNamee, P.: Language identification: a solved problem suitable for undergraduate instruction. Journal of Computing Sciences in Colleges 20(3), 94–101 (2005)
[164] Mehler, A.: Evolving lexical networks. a simulation model of terminological alignment. In: Benz, A., Ebert, C., van Rooij, R. (eds.) Proceedings of the Workshop on Language, Games, and Evolution at the 9th European Summer School in Logic, Language and Information (ESSLI 2007), Trinity College, Dublin, 6-17 August (2007)
[165] Meilă, M.: Comparing clusterings: an axiomatic view. In: Proceedings of the 22nd international conference on Machine Learning (ICML-05). pp. 577–584. ACM Press, New York, NY, USA (2005)
[166] Meilă, M., Shi, J.: Learning Segmentation by Random Walks. In: Neural Information Processing Systems (NIPS). pp. 873–879. Breckenridge, CO (USA) (2000)
[167] Merris, R.: Graph Theory. John Wiley (2001)
[168] Meyer, H.: Deutsche Sprachstatistik. Georg Olms Verlagsbuchhandlung (1967)
[169] Mihail, M., Gkantsidis, C., Saberi, A., Zegura, E.: On the semantics of internet topologies. Tech. rep., Georgia Institute of Technology (2002)
[170] Mihalcea, R.: SEMCOR semantically tagged corpus. unpublished manuscript, http://www.cse.unt.edu/ rada/downloads.htm (1998)
[171] Mihalcea, R., Chklovsky, T., Kilgarriff, A.: The SENSEVAL-3 English lexical sample task. In: Proceedings of SENSEVAL-3: Third International Workshop on the Evaluation of Systems for the Semantic Analysis of Text. pp. 25–28. New Brunswick, NJ, USA (2004)

[172] Mihalcea, R., Edmonds, P. (eds.): Senseval-3: Third International Workshop on the Evaluation of Systems for the Semantic Analysis of Text. Association for Computational Linguistics, Barcelona, Spain (July 2004)

[173] Mihalcea, R., Moldovan, D.: Automatic Generation of a Coarse Grained WordNet. In: Proceedings of the NAACL workshop on WordNet and Other Lexical Resources. Pittsburg, USA (2001)

[174] Mihalcea, R., Radev, D.: Graph-based Natural Language Processing and Information Retrieval. Cambridge University Press (June 2011)

[175] Mihalcea, R., Tarau, P.: TextRank: Bringing Order into Texts. In: Proceedings of the conference on Empirical Methods in Natural Language Processing (EMNLP-04). pp. 404–411. Barcelona, Spain (July 2004)

[176] Milgram: The small world problem. Psychology Today 1(61) (1967)

[177] Miller, G.A.: Some effects of intermittent silence. American Journal of Psychology 70, 311–313 (1957)

[178] Miller, G.A., Beckwith, R., Fellbaum, C.D., Gross, D., Miller, K.: WordNet: An On-line Lexical Database. International Journal of Lexicography 3(4) pp. 235–244 (1990)

[179] Miller, G.A., Charles, W.G.: Contextual Correlates of Semantic Similarity. Language and Cognitive Processes 6(1), 1–28 (1991)

[180] Moore, R.C.: On Log-Likelihood-Ratios and the Significance of Rare Events. In: Lin, D., Wu, D. (eds.) Proceedings of the conference on Empirical Methods in Natural Language Processing (EMNLP-04). pp. 333–340. Association for Computational Linguistics, Barcelona, Spain (July 2004)

[181] Mukherjee, A., Choudhury, M., Basu, A., Ganguly, N.: Modeling the structure and dynamics of the consonant inventories: a complex network approach. In: Proceedings of the 22nd International Conference on Computational Linguistics - Volume 1. pp. 601–608. COLING '08, Association for Computational Linguistics, Stroudsburg, PA, USA (2008)

[182] Newman, M.E.J., Watts, D.J., Strogatz, S.H.: Random graph models of social networks. Proc. Natl. Acad. Sci. USA 99(suppl. 1), 2566–2572 (February 2002)

[183] Ney, H., Essen, U., Knese, R.: On structuring probabilistic dependences in stochastic language modelling. Computer Speech and Language 8(1), 1–38 (1994)

[184] Nivre, J.: Inductive Dependency Parsing, Text, Speech and Language Technology, vol. 34. Springer (2006)

[185] Noreen, E.W.: Computer-Intensive Methods for Testing Hypotheses: An Introduction. Wiley-Interscience (1989)

[186] Norvig, P.: On Chomsky and the Two Cultures of Statistical Learning. http://norvig.com/chomsky.html [August 2011] (2011)

[187] Och, F.J.: Statistical Machine Translation: The Fabulous Present and Future. Invited talk at ACL Workshop on Building and Using Parallel Texts, Ann Arbor, Michigan, USA (2005)

[188] Olney, A.M.: Latent Semantic Grammar Induction: Context, Projectivity, and Prior Distributions. In: Proceedings of the Second Workshop on TextGraphs:

Graph-Based Algorithms for Natural Language Processing. pp. 45–52. Association for Computational Linguistics, Rochester, NY, USA (2007)
[189] Pantel, P., Lin, D.: Discovering word senses from text. In: KDD '02: Proceedings of the eighth ACM SIGKDD international conference on Knowledge discovery and data mining. pp. 613–619. ACM Press, New York, NY, USA (2002)
[190] Pantel, P., Ravichandran, D., Hovy, E.: Towards terascale knowledge acquisition. In: Proceedings of the 20th international conference on Computational Linguistics (COLING-04). p. 771. Association for Computational Linguistics, Morristown, NJ, USA (2004)
[191] Pedersen, T., Bruce, R.: Distinguishing word senses in untagged text. In: Proceedings of the Second Conference on Empirical Methods in Natural Language Processing. pp. 197–207. Providence, RI (August 1997)
[192] Peters, C. (ed.): Working notes for the CLEF 2006 Workshop. Alicante, Spain (2006)
[193] Pollard, C., Sag, I.A.: Head-Driven Phrase Structure Grammar. University of Chicago Press and CSLI Publications, Chicago, Illinois (1994)
[194] Potts, R.B.: Some Generalized Order-Disorder Transformations. Proceedings of the Cambridge Philosophical Society 48, 106–109 (1952)
[195] Pradhan, S.S., Loper, E., Dligach, D., Palmer, M.: SemEval-2007 task 17: English lexical sample, SRL and all words. In: Proceedings of the 4th International Workshop on Semantic Evaluations. pp. 87–92. SemEval '07, Association for Computational Linguistics, Stroudsburg, PA, USA (2007)
[196] Purandare, A., Pedersen, T.: Word sense discrimination by clustering contexts in vector and similarity spaces. In: Proceedings of CoNLL-2004. pp. 41–48. Boston, MA, USA (2004)
[197] Quasthoff, U., Biemann, C.: Measuring Monolinguality. In: Proceedings of the LREC-06 workshop on Quality Assurance and Quality Measurement for Language and Speech Resources. Genova, Italy (2006)
[198] Quasthoff, U., Richter, M., Biemann, C.: Corpus Portal for Search in Monolingual Corpora. In: Proceedings of the fifth international conference on Language Resources and Evaluation (LREC-06). pp. 1799–1802 (2006)
[199] Raghavan, U.N., Albert, R., Kumara, S.: Near linear time algorithm to detect community structures in large-scale networks. Physical Review E 76(3) (Sep 2007)
[200] Ramakrishnan, G., Chakrabarti, S., Paranjpe, D., Bhattacharya, P.: Is question answering an acquired skill? In: Proceedings of the 13th international conference on World Wide Web (WWW-04). pp. 111–120. ACM Press, New York, NY, USA (2004)
[201] Rapp, R.: Die Berechnung von Assoziationen: ein korpuslinguistischer Ansatz. Olms (1996)
[202] Rapp, R.: A practical solution to the problem of automatic word sense induction. In: Proceedings of the ACL 2004 on Interactive poster and demonstration sessions. pp. 26–29. Association for Computational Linguistics, Morristown, NJ, USA (2004)

[203] Rapp, R.: A Practical Solution to the Problem of Automatic Part-of-Speech Induction from Text. In: Conference Companion Volume of the 43rd Annual Meeting of the Association for Computational Linguistics (ACL-05). Ann Arbor, Michigan, USA (2005)

[204] Roget, P.M.: Roget's Thesaurus of English Words and Phrases. Penguin Books (1852)

[205] Rosenberg, A., Hirschberg, J.: V-Measure: A Conditional Entropy-Based External Cluster Evaluation Measure. In: Proceedings of the 2007 Joint Conference on Empirical Methods in Natural Language Processing and Computational Natural Language Learning (EMNLP-CoNLL). pp. 410–420 (2007)

[206] Roth, D., van den Bosch, A. (eds.): Proceedings of the Sixth Workshop on Computational Language Learning (CoNLL-02). Taipei, Taiwan (2002)

[207] Sahlgren, M.: The Word-Space Model: Using distributional analysis to represent syntagmatic and paradigmatic relations between words in high-dimensional vector spaces. Ph.D. thesis, Stockholm University (2006)

[208] Salton, G., Wong, A., Yang, C.S.: A vector space model for automatic indexing. Communications of the ACM 18(11), 613–620 (1975)

[209] Sapir, E.: Language: An introduction to the study of speech. Harcourt, Brace and company (1921)

[210] Sarkar, A., Haffari, G.: Inductive Semi-supervised Learning Methods for Natural Language Processing. Tutorial at HLT-NAACL-06, New York, USA (2006)

[211] de Saussure, F.: Course in General Linguistics. New York: McGraw-Hill (1966)

[212] Schaeffer, S.E.: Stochastic local clustering for massive graphs. In: Ho, T.B., Cheung, D., Liu, H. (eds.) Proceedings of the Ninth Pacific-Asia Conference on Knowledge Discovery and Data Mining (PAKDD-05). Lecture Notes in Computer Science, vol. 3518, pp. 354–360. Springer-Verlag GmbH, Berlin/Heidelberg, Germany (2005)

[213] Schank, T., Wagner, D.: Approximating clustering-coefficient and transitivity. Tech. Rep. 2004-9, Universität Karlsruhe, Fakultät für Informatik (2004), http://www.ubka.uni-karlsruhe.de/cgi-bin/psview?document=ira/2004/9

[214] Schank, T., Wagner, D.: Finding, Counting and Listing all Triangles in Large Graphs, An Experimental Study. In: Proceedings on the 4th International Workshop on Experimental and Efficient Algorithms (WEA-05). Lecture Notes in Computer Science, vol. 3503. Springer-Verlag (2005)

[215] Schmid, H.: Probabilistic Part-of-Speech Tagging Using Decision Trees. In: International Conference on New Methods in Language Processing. Manchester, UK (1994)

[216] Schulze, B.M.: Automatic language identification using both N-gram and word information. US Patent No. 6,167,369 (2000)

[217] Schütze, H., Pedersen, J.O.: Information Retrieval Based on Word Senses. In: Fourth Annual Symposium on Document Analysis and Information Retrieval (1995)

[218] Schütze, H.: Part-of-speech induction from scratch. In: Proceedings of the 31st annual meeting on Association for Computational Linguistics (ACL-93). pp. 251–258. Association for Computational Linguistics, Morristown, NJ, USA (1993)
[219] Schütze, H.: Word space. In: Hanson, S., Cowan, J., Giles, C. (eds.) Advances in Neural Information Processing Systems 5. Morgan Kaufmann Publishers (1993)
[220] Schütze, H.: Distributional part-of-speech tagging. In: Proceedings of the 7th Conference on European chapter of the Association for Computational Linguistics (EACL-95). pp. 141–148. Morgan Kaufmann Publishers Inc., San Francisco, CA, USA (1995)
[221] Schütze, H.: Automatic word sense discrimination. Computational Linguistics 24(1), 97–123 (1998)
[222] Shi, J., Malik, J.: Normalized Cuts and Image Segmentation. IEEE Transactions on Pattern Analysis and Machine Intelligence 22(8), 888–905 (2000)
[223] Sigurd, B., Eeg-Olofsson, M., van de Weijer, J.: Word length, sentence length and frequency – Zipf revisited. Studia Linguistica 58(1), 37–52 (2004)
[224] Šíma, J., Schaeffer, S.E.: On the NP-Completeness of Some Graph Cluster Measures. Technical Report cs.CC/0506100, arXiv.org e-Print archive (June 2005)
[225] Simon, H.A.: On a class of skew distribution functions. Biometrika 42(3/4), 425–440 (1955)
[226] Smith, F.J., Devine, K.: Storing and retrieving word phrases. Inf. Process. Manage. 21(3), 215–224 (1985)
[227] Søgaard, A.: Semisupervised condensed nearest neighbor for part-of-speech tagging. In: Proceedings of the 49th Annual Meeting of the Association for Computational Linguistics: Human Language Technologies: short papers - Volume 2. pp. 48–52. HLT '11, Association for Computational Linguistics, Stroudsburg, PA, USA (2011)
[228] Steels, L.: Evolving Grounded Communication for Robots. Trends in Cognitive Sciences 7(7), 308–312 (7 2003)
[229] Stefanowitsch, A., Gries, S.T.: Collostructions: Investigating the interaction of words and constructions. International Journal of Corpus Linguistics 8, 209–243 (2003)
[230] Stein, B., Niggemann, O.: On the Nature of Structure and Its Identification. In: WG '99: Proceedings of the 25th International Workshop on Graph-Theoretic Concepts in Computer Science. pp. 122–134. Springer-Verlag, London, UK (1999)
[231] Steinbach, M., Karypis, G., Kumar, V.: A Comparison of Document Clustering Techniques. In: KDD Workshop on Text Mining. Boston, MA, USA (2000)
[232] Steyvers, M., Tenenbaum, J.B.: The Large-Scale Structure of Semantic Networks: Statistical Analyses and a Model of Semantic Growth. Cognitive Science 29(1), 41–78 (2005)

[233] Strogatz, S.: Exploring complex networks. Nature 410(6825), 268–276 (March 2001)
[234] Teresniak, S.: Statistikbasierte Sprachenidentifikation auf Satzbasis. Bachelor's thesis, University of Leipzig (2005)
[235] Thelen, M., Riloff, E.: A bootstrapping method for learning semantic lexicons using extraction pattern contexts. In: Proceedings of the ACL-02 conference on Empirical Methods in Natural Language Processing (EMNLP-02). pp. 214–221. Association for Computational Linguistics, Morristown, NJ, USA (2002)
[236] Tjong Kim Sang, E., Buchholz, S.: Introduction to the CoNLL-2000 Shared Task: Chunking. In: Proceedings of CoNLL-2000. Lisbon, Portugal (2000)
[237] Toutanova, K., Johnson, M.: A Bayesian LDA-based model for semi-supervised part-of-speech tagging. In: NIPS. pp. 1521–1528 (2007)
[238] Turney, P.D.: Similarity of semantic relations. Computational Linguistics 32(3), 379–416 (2006)
[239] Van Rijsbergen, C.J.: Information Retrieval, 2nd edition. Dept. of Computer Science, University of Glasgow (1979)
[240] Veronis, J.: Hyperlex: lexical cartography for information retrieval. Computer Speech & Language 18(3), 223–252 (July 2004)
[241] Voorhees, E.M.: Using WordNet to disambiguate word senses for text retrieval. In: SIGIR '93: Proceedings of the 16th annual international ACM SIGIR conference on Research and development in information retrieval. pp. 171–180. ACM Press, New York, NY, USA (1993)
[242] Voss, J.: Measuring Wikipedia. In: Ingwersen, P., Larsen, B. (eds.) ISSI2005. vol. 1, pp. 221–231. International Society for Scientometrics and Informetrics, Stockholm (2005)
[243] Wallach, H.M.: Topic modeling: beyond bag-of-words. In: NIPS 2005 Workshop on Bayesian Methods for Natural Language Processing (2005)
[244] Watts, D.J., Strogatz, S.H.: Collective dynamics of 'small-world' networks. Nature 393(6684), 440–442 (June 1998)
[245] Webb, G.I., Boughton, J.R., Wang, Z.: Not so naïve Bayes: Aggregating one-dependence estimators. In: Machine Learning. pp. 5–24 (2005)
[246] Wei, Y.C., Cheng, C.K.: Ratio cut partitioning for hierarchical designs. IEEE Trans. on CAD of Integrated Circuits and Systems 10(7), 911–921 (1991)
[247] Widdows, D., Dorow, B.: A graph model for unsupervised lexical acquisition. In: Proceedings of the 19th international conference on Computational Linguistics (COLING-02). pp. 1–7. Association for Computational Linguistics, Morristown, NJ, USA (2002)
[248] Witschel, H., Biemann, C.: Rigorous Dimensionality Reduction through Linguistically Motivated Feature Selection for Text Categorization. In: Proceedings of NODALIDA'05. Joensuu, Finland (2005)
[249] Wu, Z., Leahy, R.: An Optimal Graph Theoretic Approach to Data Clustering: Theory and Its Application to Image Segmentation. IEEE Trans. Pattern Anal. Mach. Intell. 15(11), 1101–1113 (1993)

[250] Yarowsky, D.: Decision lists for lexical ambiguity resolution: application to accent restoration in Spanish and French. In: Proceedings of the 32nd annual meeting on Association for Computational Linguistics (ACL-94). pp. 88–95. Las Cruces, New Mexico (1994)
[251] Yarowsky, D.: Unsupervised word sense disambiguation rivaling supervised methods. In: Proceedings of the 33rd Annual Meeting of the Association for Computational Linguistics (ACL-95). pp. 189–196. Cambridge, MA (1995)
[252] Zanette, D.H., Montemurro, M.A.: Dynamics of Text Generation with Realistic Zipf's Distribution. Journal of Quantitative Linguistics 12(1), 29–40 (2005)
[253] Zha, H., He, X., Ding, C., Simon, H., Gu, M.: Bipartite graph partitioning and data clustering. In: CIKM '01: Proceedings of the tenth International Conference on Information and Knowledge Management. pp. 25–32. ACM Press, New York, NY, USA (2001)
[254] Zhu, X.: Semi-Supervised Learning Literature Survey. Tech. rep., Computer Sciences, University of Wisconsin-Madison (2005)
[255] Zipf, G.K.: The Psycho-Biology of Language. Houghton Mifflin, Boston (1935)
[256] Zipf, G.K.: Human Behavior and the Principle of Least-Effort. Addison-Wesley, Cambridge, MA (1949)

Printed by Publishers' Graphics LLC USA
MO20120327-125
2012